Crop Modelling for Land Use Planning

Crop Modelling for Land Use Planning

Editor

Kishor Gore

scitus
academics

Crop Modelling for Land Use Planning

Edited by **Kishor Gore**

Printed in 2017

ISBN: 978-1-68117-055-8

Library of Congress Control Number: 2015935453

SCITUS Academics LLC,
616, Corporate Way, Suite 2, 4766,
Valley Cottage, NY 10989

www.scitusacademics.com

Contents

Preface

Current land use issues, which require a resolution formulated with the aid of this approach in the rural and peri-urban spheres, are frequently derived from environmental versus developmental conflicts. Those discussed include decision making on whether it is preferable to use scarce resources to rehabilitate degraded land or to improve prime agricultural land, whether smallholder settlements or large-scale mechanized farming will better support the expanding population, the encroachment of urban development onto high quality agricultural land, the correct uses of scarce water resources, and the particular requirement for integrated as opposed to sectoral planning of coastal zones. The execution of the integrated approach, will depend on policies that support planning for the use and sustainable management of land resources, on the strengthening of implementing institutions and on ensuring the active involvement and participation of stakeholders in the decision-making process. These actions will in turn be supported by a variety of databases on natural resources and their uses, which are combined through the use of a geographical information system. Social and economic tools are also described, which when used will ensure the inclusion of the contributions from stakeholders in land use negotiations. An integrated approach to planning the use and management of land resources entails the involvement of all stakeholders in the process of decision making on the future of the land, and the identification and evaluation of all biophysical and socio-economic attributes of land units. This requires the identification and establishment of a use or non-use of each land unit that is technically appropriate, economically viable, socially acceptable and environmentally non-degrading.

Editor

Chapter 1

Modeling Biofuel Expansion Effects on Land Use Change Dynamics

Ethan Warner[1], Daniel Inman[1], Benjamin Kunstman[1], Brian Bush[1], Laura Vimmerstedt[1], Steve Peterson[2,3,4], Jordan Macknick[1], and Yimin Zhang[1]

[1]National Renewable Energy Laboratory, 15013 Denver West Parkway, Golden, CO, USA

[2]Peterson Group, Lebanon, NH, USA

[3]Lexidyne LLC, Colorado Springs, CO, USA

[4]Thayer School of Engineering, Dartmouth College, Hanover, NH, USA

ABSTRACT

Increasing demand for crop-based biofuels, in addition to other human drivers of land use, induces direct and indirect land use changes (LUC). Our system dynamics tool is intended to complement existing LUC modeling approaches and to improve the understanding of global

LUC drivers and dynamics by allowing examination of global LUC under diverse scenarios and varying model assumptions. We report on a small subset of such analyses. This model provides insights into the drivers and dynamic interactions of LUC (e.g., dietary choices and biofuel policy) and is not intended to assert improvement in numerical results relative to other works.

Demand for food commodities are mostly met in high food and high crop-based biofuel demand scenarios, but cropland must expand substantially. Meeting roughly 25% of global transportation fuel demand by 2050 with biofuels requires >2 times the land used to meet food demands under a presumed 40% increase in per capita food demand. In comparison, the high food demand scenario requires greater pastureland for meat production, leading to larger overall expansion into forest and grassland. Our results indicate that, in all scenarios, there is a potential for supply shortfalls, and associated upward pressure on prices, of food commodities requiring higher land use intensity (e.g., beef) which biofuels could exacerbate.

INTRODUCTION

Biofuel production has been pursued because of opportunities to contribute to climate change mitigation, among other potential benefits such as securing and diversifying energy supply and providing, economic development opportunities especially in rural areas. Initial biofuel assessments (e.g., Farrell *et al* 2006) suggested that biofuels, such as corn ethanol, could help the United States (US) reduce greenhouse gas (GHG) emissions. However, Searchinger *et al* (2008) and Fargione *et al* (2008) highlighted that previous biofuel studies failed to include the effects of global land use change (LUC). These two watershed studies modeled the potential impact of carbon released from soils and above-ground biomass during land clearing activities triggered by increased demand for biofuel. Results from these studies, as well as many other biofuel-induced LUC studies (e.g., see Berndes *et al* 2010), hinge on assumptions regarding direct and indirect causal relationships (see section 1 of the supplemental information (SI), available at stacks.iop.org/ERL/8/015003/mmedia, for more detail) between drivers of LUC, land availability, biomass yields, population dynamics, dietary choices, relative affluence, and biofuel demand, to name a few.

However, the causal relationships that underpin such results are not completely understood and there remains significant disagreement on many fundamental aspects of LUC dynamics. LUC, as well as those changes attributable to biofuels, has far-reaching implications for many aspects of sustainability i.e., biodiversity and societal impacts (e.g., food security). For example, diversion of land to biofuel crops displaces production that may lead to compensating production of substitute crops elsewhere, affecting regional food crop prices (Chum *et al*2011, Dale *et al* 2011). Therefore, a better understanding of the drivers of LUC and high-level influences on system behavior is critical to the responsible and sustainable development of biofuels.

Inclusion of LUC impacts in renewable fuels policy is contentious, because it is neither directly measurable nor easily isolated from the myriad of other LUC drivers (Plevin *et al* 2010), such as agricultural policies, agricultural product demand changes, and social norms. Biofuel policy analyses typically rely on computer simulations or on extrapolations of historic data to evaluate total LUC. The LUC modeling science lacks consensus with regard to modeling frameworks, boundary conditions, and other fundamental assumptions, which has resulted in highly variable modeling results across a wide range of studies. For example, results of CO_2emissions from biofuel-induced LUC span an order of magnitude, and subsequent calculations of GHG emissions can even vary in sign (Berndes *et al* 2010). Currently, US and European governmental organizations are integrating—or are considering integrating—LUC impacts into their renewable fuel policies (e.g., the US Renewable Fuel Standard (RFS) US EPA 2010), EU Renewable Energy and Fuel Quality Directive (European Commission 2009), United Kingdom (UK) Renewable Transport Fuel Obligation (Gallagher 2008, UK Department for Transport 2012), and California's Low Carbon Fuel Standard (CARB 2009). At present, analysts often use agricultural economic models. For example, variations of the Global Trade Analysis Project (GTAP) database and the US EPA's methodologies are commonly being used to estimate GHG_{LUC} for renewable fuel policy purposes (US EPA2010, Tyner *et al* 2010, Al-Riffai *et al* 2010).

The modeling approach presented in this study is intended to complement existing approaches and to improve the understanding of global LUC drivers and dynamics. Every effort has been made to ensure the model is parsimonious and transparent, both in terms of the underlying data and the feedback effects among drivers. By using a model with high transparency, ease of use, and dynamic capabilities, our study improves policy-relevant analyses by allowing examination of global LUC under diverse scenarios and varying model assumptions. In this paper, we report on a small subset of such analyses. These are intended as an initial illustration of model functionality, not to assert improvement in numerical results relative to other work.

LUC MODELS AND APPROACHES

Based on our review of the literature, we categorize most biofuel-related LUC modeling approaches into three broad categories: (1) general and partial economic equilibrium modeling, (2) simple deterministic methods, and (3) causal descriptive methods. Each method has strengths and weakness in modeling LUC and its causes (see table 1), because purposes of models vary. Modeling classifications listed in table 1 suggest a spectrum of complexity, with higher complexity on the left side.

Table 1: Comparison of models using different methods used to estimate LUC in existing literature (Ackerman 2002, ICF International 2009, CBES 2009). Specific models may differ from these generalizations

	General and partial economic equilibrium models	System dynamics	Causal descriptive method	Deterministic method (simplified)
Description	Calculates LUC resulting from the cause-and-effect relationships in a market system based on the change resulting from shocking an economic system at equilibrium with an expansion of biofuels. Uses historic data for calibration, but may incorporate empirically estimated functions. Model parameters are often derived from related literature	Calculates LUC using cause-and-effect relationships in a dynamic stock-and-flow framework to assess market responses resulting from increasing demand for biofuels. Uses historic trends as inputs and for calibrating cause-and-effect relationships	Calculates LUC using cause-and-effect relationships to assess market responses resulting from increasing demand for biofuels. Uses historic trends as inputs and for calibrating cause-and-effect relationships	Generates an 'average' adder for agricultural products considered relevant to the bioenergy sector based on historic land use data. Direct and indirect LUC are sometimes combined but sometimes separate factors

Potentially limiting assumptions in existing models	(1) LUC is driven primarily by price-related economic forces. (2) Non-price factors are assumed to be exogenous and generally held constant	(1) Future market responses (e.g., product substitution, cultivation area of a given crop) to increased demand for biofuels can be extrapolated from historical trends and relationships between economic, agricultural, and social systems. (2) Market responses are driven by relatively independent external assumptions, but internal moderating feedbacks are included	(1) Future market responses (e.g., product substitution, cultivation area of a given crop) to increased demand for biofuels can be extrapolated from statistical analysis of historical trends. (2) Market responses are relatively independent and can be quantified without considering possible correlations	(1) Only countries trading evaluated agricultural commodities might be subject to LUC effects. (2) The patterns of global trade of agricultural commodities will remain virtually unchanged in the near future, and therefore the incremental effect of biofuel expansion on land use will always have the same global impact. (3) LUC caused by increased biofuel feedstock production can be directly estimated from the type and share of land used to grow agricultural commodities for export purpose

Strengths	(1) Captures, in detail, many important market and economic factors that can drive LUC. (2) Usually based at institutions that offers technical support. (3) In its dynamic form can account for the evolution of demographic variables and resources, and other changes over time	(1) Can relatively easily incorporate important LUC drivers other than market and economic factors. (2) Can be transparent and involve stakeholders in running the model and scenario development	(1) Can relatively easily incorporate important LUC drivers other than market and economic factors. (2) Can be transparent and involve stakeholders in running the model and scenario development	(1) Easy to calculate and transparent
Limitations	(1) Currently, usually fails to include other important LUC drivers (e.g., political, cultural, demographic, environmental forces) that may not rely on land and commodity prices and elasticities. (2) Difficult to gain access to or use by non-experts economic modelers	(1) Misses dynamic and interlinked LUC driving factors in cause-and-effect chains because of reliance on historical data and relationships to identify LUC drivers. (2) Data and relationships operate at a relatively aggregate low-resolution level. (3) Prices need not be explicitly modeled	(1) Misses dynamic and interlinked LUC driving factors in cause-and-effect chains because of reliance on historical data and relationships to identify LUC drivers. (2) Data and relationships operate at a relatively aggregate low-resolution level. (3) Prices need not be explicitly modeled	(1) Generally does not consider land types not used for producing traded agricultural commodities. (2) Omits market feedbacks and policy measures that affect trade patterns. (3) Data and relationships operate at a very aggregate low-resolution level

Example papers	Hiederer et al (2010), Al-Riffai et al (2010), US EPA (2010), Tyneret al (2010), Wise et al(2009)	Sheehan (2009)	Bauen et al (2010), Lywood (2008)	Kim et al (2009), Fritsche et al(2010), Tipper et al (2009)

The new model, called BioLUC, is a system dynamics (SD) simulation model (NREL 2012, Bush *et al* 2011) that represents key economic and social drivers of global LUC and their interactions over time, enabling exploration of different scenarios with implications for LUC[5]. In particular, BioLUC can explore implications of and assumptions about LUC by analyzing the limits of sustainable biofuels production under varied future conditions regarding inputs, specifically population growth, crop yields, and plant and animal product supply and demand. BioLUC was created using the STELLA Version 9.1.4 software package (ISEE Systems, Lebanon, NH) using a stock-and-flow structure; it focuses on information feedback processes that underwrite the dynamic movement of key quantities over time.

Footnote

[5]The authors are willing to provide a copy of the model, model internal equations or related milestone reports on the model, as requested. Official public release of the model will not be complete until 2013.

BioLUC generally would fit the causal descriptive category. However, to-date, causal descriptive modeling systems have been spreadsheet-based models that lack dynamic stock-and-flow frameworks. To explain the distinctions we describe SD models in a separate column in table 1. See section 2 of the SI (available at stacks.iop.org/ERL/8/015003/mmedia) for additional details about SD modeling, its historical context, and its use in projecting the consequences of today's and potential future policies.

The BioLUC model recognizes the accomplishments of past LUC modeling efforts and provides a modeling option that may address some of the limitations of current methods evaluating LUC and complement and existing suite of models to improve research insights. Transparency and ease of use are challenging to achieve when examining issues of biofuel-induced LUC, especially for the many interested stakeholders who are not economic modeling specialists. Enabling this community to access a transparent analytic method can help them work together to understand and analyze underlying LUC drivers; test assumptions about LUC systems against historic data; investigate future conditions; and assess implications of new LUC research results (Sheehan 2009, 2012). As tools for this kind of shared exploration, existing models are either too complex, with light model structure documentation, difficult data access, or, alternatively, grossly simplistic (i.e., deterministic method). A SD modeling approach building on the causal descriptive method attempts to fill this analysis space.

BIOLUC MODELING APPROACH

The generalized influence diagram for a generic region of the world in the BioLUC model is presented in figure 1. The movement of land between usage categories over time is represented in the model using stocks for four different land bases, which flows into or out of those stocks. Within the 'crops' land base, land is allocated among multiple competing uses (e.g., food, feed, fuel, and fiber). Note the 'abandoned' category among the land bases: this land category, assumed nonproductive, enables us to explore land abandonment and rehabilitation scenarios. 'Available' land is forest and grassland that is potentially available for productive use as pasture or cropland.

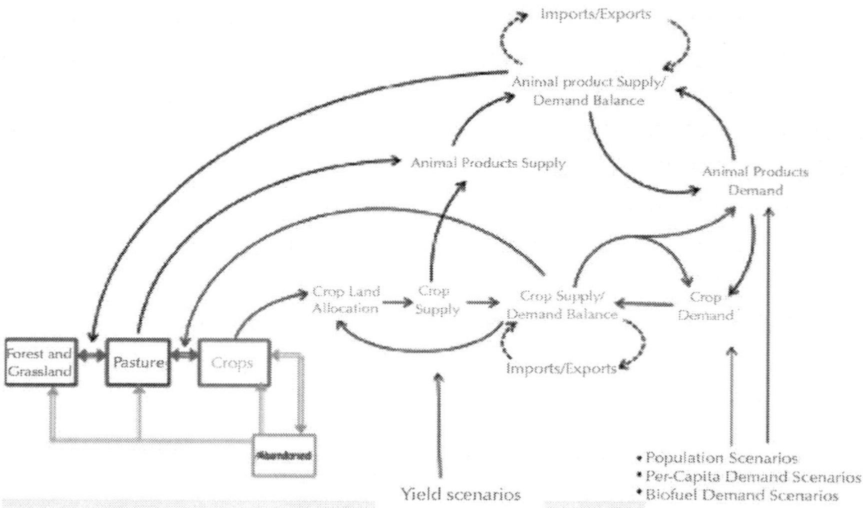

Figure 1: also shows how externally defined scenarios for key parameters impact the system: biomass yield and crop land allocation determine production of various crops; and population, per capita demand, and biofuel demand drives 'direct' demands for various crops.

BioLUC represents key feedback processes that drive the allocation of land over time. Examples of these processes include:

- Imbalances between production and consumption of various agricultural products motivate changes in the allocation of land among different uses at a regional level.
    1. Demand for animal products creates additional demand for crops grown as feed.
    2. Crop and animal product imbalances between production and consumption stimulate adaptive responses in the system to move toward equilibrium.
    3. Reallocation of existing crop land among different uses balances the mix of crops produced against the mix of crops required.
- Re-distribution of the land bases, for example by converting available land into pastureland or by turning pastureland into crop land, adjusts production to more closely meet demand.
- Crop or animal product imbalances are further reduced through imports/exports from other regions.

An imbalance between demand and supply stimulates multiple feedback mechanisms. For example, holding other things equal, if the combination of population, per capita demand, and biofuel scenarios cause consumption of a particular crop to exceed its production within a region several processes will begin to unfold:

- Regional inventories of the crop will begin to decline.
- The resulting supply shortfall will constrain consumption to levels lower than those implied by population, per capita demand and biofuel scenarios.
- The supply shortfall will cause the region to call for imports from outside the region, in the immediate term.
- Additionally, the supply shortfall will lead to a reallocation of crop land in favor of the crop in question in the longer term.
- Reallocation of cropland will increase the rate at which land moves from pasture into crops, which in turn will increase pressure to convert land from forest or grassland into pasture.

As these processes play out over time, the system will seek to balance itself so that equilibrium between supply and demand for crops within a region is restored.

We initially developed a two-region model that can be used to represent any two regions. See section 2 of the SI (available atstacks.iop.org/ERL/8/015003/mmedia) for more detail on two-region model structure and intra-regional trade. For additional details about model input assumptions including agricultural commodity yields, changes in population, and initial land cover at the start of the model across all scenarios, see data and model calculations in section 3 of the SI (available atstacks.iop.org/ERL/8/015003/mmedia).

BIOLUC MODEL SCENARIOS

We explore four scenarios to broadly examine the effects of demand for crop-based biofuels and food on LUC as outlined below and in table 2. We constructed scenarios to represent an extreme high-intensity agricultural future to adequately highlight model dynamics and test model integrity under high pressures. The details of these scenarios are included in section 4 of the SI (available at stacks.iop.org/ERL/8/015003/mmedia), which describes biofuel yields, conversion yields, food and feed demand, and biofuel demand assumptions for several time steps of each scenario.

Table 2: BioLUC demand scenarios. The scenarios are modeled from 1990–2050. Input data is annual, but the model runs on a time step of 1/32nd of a year

Conditions in 2000	• 9 million ha harvested for biofuels	
	• 0.6 EJ biofuels	
	• 110–280 kg meat and dairy/capita-yr	
	• 260–300 kg other food/capita-yr	
	Lower biofuels demand	Higher biofuels demand
Lower food demand	Business-as-usual (BAU) scenario	Higher biofuel (HB) scenario
	By 2050:	By 2050:
	• 80 million ha harvested for biofuels	• 700 million ha harvested for biofuels
	• 7 EJ biofuels	• 46 EJ biofuels
	• 140–310 kg meat and dairy/capita-yr	• 140–310 kg meat and dairy/capita-yr
	• 280–290 kg other food/ capita-yr	• 280–290 kg other food/ capita-yr
Higher food demand	Higher food (HF) scenario	Higher food and biofuel (HFB) scenario
	By 2050:	By 2050:
	• 80 million ha harvested for biofuels	• 700 million ha harvested for biofuels
	• 7 EJ biofuels	• 46 EJ biofuels
	• 200–360 kg meat and dairy/capita-yr	• 200–360 kg meat and dairy/capita-yr
	• 330–340 kg other food/ capita-yr	• 330–340 kg other food/ capita-yr

The two crop-based biofuel demand levels are presumed to be policy driven rather than economically driven. The basis for setting the demand levels is as follows:

- Lower biofuels demand is mostly based on levels and assumptions from Alexandratos and Bruinsma (2012), in which current (as of 2012) biofuels policies are met through increased use of food-based crops to 2020, with no subsequent growth. One adjustment made to Alexandratos and Bruinsma (2012) was the use of cellulosic ethanol starting in 2016 to meet ethanol requirements. Corn ethanol was effectively capped at the policy mandated level of 57 billion dm3 (15 billion gallons) (US EPA 2010).
- Higher biofuels demand based on linear growth to reach a 25% global displacement of gasoline and diesel by 2050 using advanced (cellulosic and renewable diesel) biofuel systems, in addition to the biofuels in the lower biofuel demand scenario. Displacement of fossil fuels is calculated on an energy basis, with no petroleum market rebound effect.The two food and feed demand levels are:
- Lower demand growth based on levels of per capita demand for food from Alexandratos and Bruinsma (2012).
- Higher demand growth based on high-end demand projections (~40% increases in per capita food demand by 2050 from 2005) were taken from Tilman et al (2011) and modified to be applied to the aforementioned projections from Alexandratos and Bruinsma (2012). Tilman et al (2011) does not specify how the demand increase is distributed across individual food categories. We closely approximate the ~40% increase by equally applying changes across all commodities through a 45% increase in annual growth of each commodity demanded starting in 2010 relative to the low food demand scenario.

The four scenarios, presented in table 2, are limited in at least two key respects.

First, our examination of yield is constrained. FAO (2010) resolution is limited to national-level averages. In our high demand scenarios, we do not assume higher levels of agricultural intensification (i.e., higher than BAU yields increases) in response to economic forces, nor do we examine reductions in crop yields, as could result from climate change and increases in extreme weather events. Our yield data (from FAOSTAT) are aggregate national averages that can have large internal spatial and temporal variability. A limited scenario analysis examining the impact of higher and lower cellulosic biofuel yield trend assumptions was examined in section 6 of the SI (available at stacks.iop.org/ERL/8/015003/mmedia) and is discussed briefly in the results section, below.

Second, we model the high biofuels case essentially as higher land requirements to grow the biofuel feedstock on agricultural land (i.e., not wastes or residues or grown on marginal lands). We selected these limitations to test extreme land use conditions and to simplify the analyzed scenarios. The suite of feedstocks grown and technologies used is essentially generic: a shift in technology, feedstock, or yield assumption would only alter the aggregate land requirements (i.e., ~700 million ha) examined in this analysis. The implications of these limitations will be discussed further in our results.

RESULTS AND DISCUSSION

BioLUC results are not predictions; the model only provides insights into the drivers and dynamic interactions of LUC. Quantities and changes are only provided to facilitate comparisons between our scenarios.

Business-as-usual (BAU) projections of LUC for the US and ROW are presented in figure 3. These results reflect changes in land use in response to global population growth and an overall increase in per capita gross domestic product, which causes diets to shift toward more calories per capita as well as a greater percentage of those calories coming from meat and animal products (e.g., dairy) (Alexandratos and Bruinsma 2012). Results suggest that the rate of conversion of available land (i.e., forest and grasslands) to cropland increases globally, relative to historic data in the US and the ROW, in order to meet rising food needs. In the ROW the rate of cropland increase remains similar to historic rates of change, but pastureland grows significantly. Pastureland trends in the ROW reflect the larger relative shift in per capita gross domestic product in developing countries. Land conversion begins to slow circa 2030 because the rate of population growth and per capita food demand growth begins to stabilize.

Globally, the most noticeable trend in the ROW is the increase in pastureland area to meet meat and animal product demands as the world's population grows and becomes more affluent. This is consistent with well-documented historic trends that show that increased wealth prompts a shift from diets rich in whole vegetables toward diets that have greater amounts of processed grains and more meat and animal products (Southgate *et al* 2007). The type of meat consumed matters: in general, the larger the animal, the greater the ratio of biomass to animal mass. Transitioning to eating more beef (and other larger animals' meat) would require even larger amounts of land than is predicted in the BAU scenario. Alexandratos and Bruinsma (2012) assume that protein consumption will increase in the developing world, but consumption levels will be lower and the commodity mix will be different than has been seen in the developed world (e.g., the US).

Our BAU scenario assumes 2050 biofuel energy requirements will be the same as in 2020, and results in an estimate of 80 million ha of land globally needed to be used for energy crops to meet those requirements in 2050. None of the four scenarios considered in our analysis uses residues and wastes, which would have negligible land use effects (US EPA 2010). For example, biofuel production from forestry residues, agricultural residues, and wastes could supply about another 5% (9 EJ yr^{-1}) of global transportation fuel if these resources were allocated to cellulosic ethanol production, based on even the most pessimistic technical potential assumptions from Chum *et al* (2011). Another biofuel-related limitation is that our available land stock (and pastureland and cropland stocks) contains land of a wide variety of qualities. If cellulosic crops use abandoned or less productive lands, the impact on prime agriculture lands may be lessened (US EPA 2010).

Land moves to cropland from pastureland first, and then from available land (i.e., forest and grassland) as shown in the HF scenario in figure 3. Table 3 lists the per cent change in land use between the HF and other alternative scenarios and the BAU scenario. In the HF scenario, cropland increases by about 15% globally by the end of the simulation in order to meet growing demand for food products compared to BAU. US cropland expansion to meet rising food requirements occurs mostly at the expense of available forest and grassland but also involves some pastureland. In the ROW, cropland expansion occurs almost exclusively on forest and grassland land. Pastureland in the ROW increases to help supply higher meat requirements. A similar dynamic is not observed in the US because high-land-use intensity meat consumption relative to other meat commodities is projected to decline (Alexandratos and Bruinsma 2012). These trends are offset in the HF scenario, but the results are a static rather than a growing consumption of high-land-use intensity meat as seen in ROW. Alexandratos and Bruinsma (2012) modeled diet trends based on historical data and dietary trajectories of other meat-consuming developed countries.

Cropland expands more significantly in the HB scenario than in the HF scenario to meet the high demand for biofuels and there are different land use tradeoffs. Globally, about 700 million ha of land are required to meet the high biofuel requirements. In the HB scenario (table 3), US cropland area increases by about 40% and the ROW cropland area by about 25% by the end of the simulation, compared to BAU. In the HB scenario, US cropland expands onto pastureland, and also available land to pastureland to compensate. The ROW cropland expansion follows this same trend, so compared to the HF scenario, much less available land is actually used.

Table 3: Per cent change in land use from BAU scenario

Scenario	Region	Land type	2020 (%)	2030 (%)	2040 (%)	2050 (%)
HF	USA	Cropland	7	12	16	18
HF	ROW	Cropland	5	9	13	16
HF	World	Cropland	5	10	14	16
HF	USA	Pastureland	−1	−5	−8	−10
HF	ROW	Pastureland	5	5	3	2
HF	World	Pastureland	4	4	2	1
HF	USA	Forest and grassland	−8	−15	−21	−27
HF	ROW	Forest and grassland	−14	−24	−32	−40
HF	World	Forest and grassland	−13	−23	−31	−39
HB	USA	Cropland	1	17	31	40
HB	ROW	Cropland	0	8	17	24
HB	World	Cropland	0	9	19	26
HB	USA	Pastureland	0	−13	−24	−32
HB	ROW	Pastureland	0	−3	−6	−8
HB	World	Pastureland	0	−3	−7	−9

HB	USA	Forest and grassland	0	−6	−14	−22
HB	ROW	Forest and grassland	0	−1	−4	−7
HB	World	Forest and grassland	0	−2	−5	−8
HFB	USA	Cropland	8	29	46	57
HFB	ROW	Cropland	5	18	30	39
HFB	World	Cropland	6	19	32	41
HFB	USA	Pastureland	−2	−18	−32	−44
HFB	ROW	Pastureland	4	2	−3	−6
HFB	World	Pastureland	4	1	−5	−9
HFB	USA	Forest and grassland	−8	−20	−30	−39
HFB	ROW	Forest and grassland	−14	−25	−34	−42
HFB	World	Forest and grassland	−13	−24	−33	−42

The HFB scenario requires expansion into available land and pastureland to accommodate the combined food and biofuel demand, as shown in figure 2. The basic trends and dynamics demonstrated in the high biofuel and high food demand scenario are extended in the combined high food and biofuel demand scenario:

- Higher food or fuel demand recruits available land for agricultural production.
- Relative changes in pastureland versus crop land reflect changes in the relative demand for meat versus crops, as well as changes in the type of meat demanded.
- Relative changes in pastureland versus available land reflect food (i.e., meat) versus fuel demand, as well as competition over the use of pastureland.
- Pastureland in the ROW remains much higher than in the US because of a more dramatic shift and continued growth in requirements for meat in the ROW.

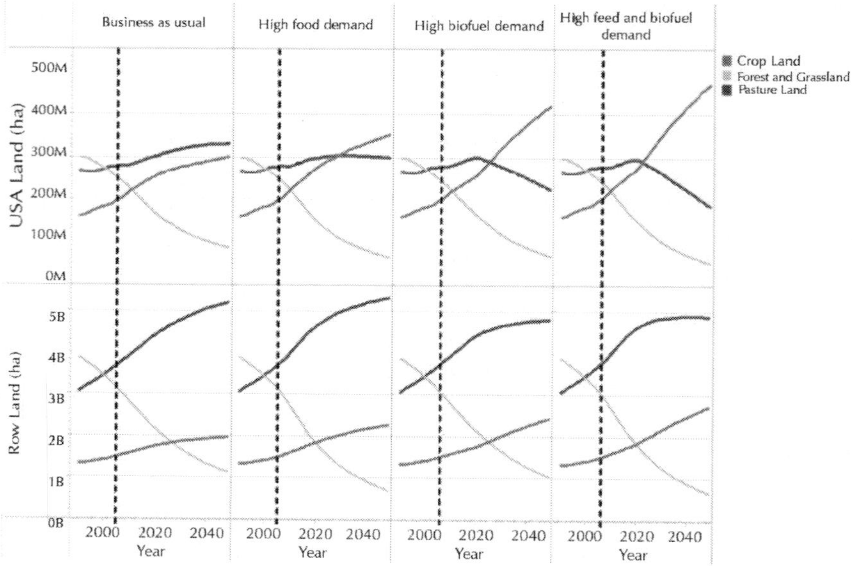

Figure 2: LUC for two regions across four commodity demand scenarios. In the BAU and high food demand scenarios, biofuel policies that are in place in as of 2012 are assumed to be met in 2019 by food-based crops. Division in historic and modeled data indicated

by dashed vertical line.

Land use change across all scenarios examined in this study is presented in figure 3. Even though pastureland increased slightly in the high food demand scenario, it decreased in the HF and HFB scenarios. These results highlight an underlying dynamic occurring, to some extent across all scenarios of the model. In some years, meat production could fall short, in particular for land-intensive commodities (e.g., beef). Demands for other commodities, such as cereals, are being met at most points in time in non-HFB scenarios. The exception is in the ROW HFB scenario between about 2010 and 2030, when the most rapid shifts are occurring in land use and commodities demanded. The rate of land conversion needed to meet increases in food and biofuel demands between 2010 and 2030 is the highest during this time period. During this time period there are unmet commodity demands (1%–3% of total) occurring every few years as the model seeks to achieve equilibrium under highly stressful conditions. After this time period, non-meat demands were accommodated.

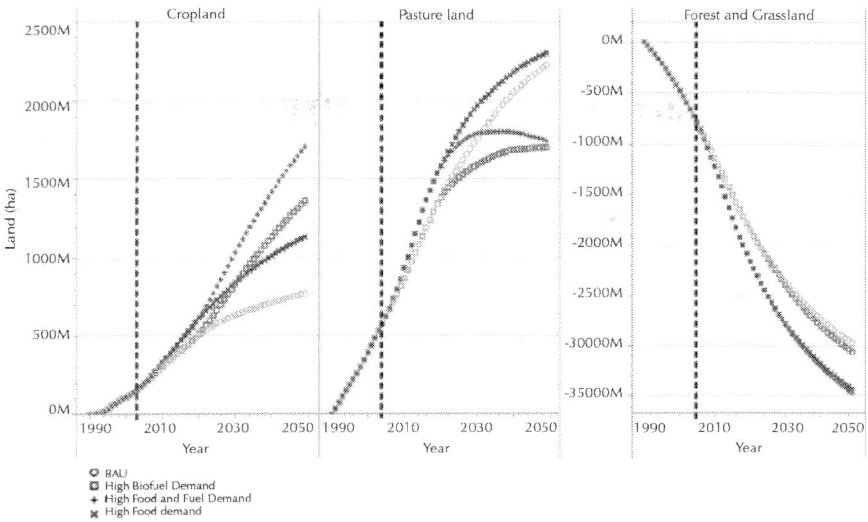

Figure 3: Global change from 1990 to 2050 in cropland, pastureland, and available land (i.e. forest and grassland), in response changes in demand. Division in historic and modeled data indicated by dashed vertical line.

As described in the methods, the model is equilibrium seeking, responding to demands for commodities. Consumption is constrained not to exceed physically and logistically feasible supply. Within a given region and time period, supply shortfalls of commodity crops lead to constraints in animal product consumption before constraining direct human consumption, so that if humans demanded both more grain and more meat, land would first be used to produce more grain. Shortfalls in practice imply that people eat less of a commodity or that if possible they shift consumption to a substitute. However, the model does not explicitly capture this potential substitution effects between animal and non-animal products with regard caloric intake. It is assumed that as prices for meat products go up people do not shift their diets toward alternative food commodities in addition to any demand reductions.

Supply shortfalls are more common in biofuel scenarios because biofuels' use of land was given priority as their use is policy driven rather than economically driven. Despite higher food demands in the HFB scenario, pastureland growth is much lower than that in the HF scenario, because land is first allocated to biofuel feedstocks. In practical terms, such shortfalls could be avoided through various means, such as by removing biofuel production requirements, structuring biofuel policies to respond to market conditions, switching diets to lower-land-use intensity meat use, or improving biofuel and food commodity. Section 5 of the SI (available atstacks.iop.org/ERL/8/015003/mmedia) gives an example future in which cellulosic biofuel yields improve or exacerbate shortfalls. An increase or decrease of 0.5% of annual average cellulosic feedstock yield growth from 2020–2050 had enough of an impact to free up or occupy substantial amounts of land. Specifically, in comparison to baseline assumptions and using Alexandratos and Bruinsma (2012), 0.4–0.5 billion people per year would or would not, have wheat demands met, depending on this range of yield growth. A negative trend in annual yield improvement due to climate change (e.g., extreme weather events) would lead to similar if more extreme results (Lobell and Christopher 2007). While yield levels may not change underlying dynamics, they do have an important role to play in the magnitude of land used by biofuels and other human uses of land.

BioLUC results are not predictions, but they may indicate when and why stresses arise in the global agricultural system. BioLUC is a simple model that focuses on bookkeeping for land stocks, food inventory, and international trade. It tends toward equilibrium conditions that allocate resources to meet demand. The model's main use is to develop insights into the interplay among the myriad of factors impinging on the global land system. Therefore, we examined the dynamics of scenarios, rather than conducting detailed uncertainty analysis around any given scenario.

We have already outlined several limitations of our analysis, but one significant limitation is the resolution of global regions. To address this, the flexible design of the BioLUC modeling framework allows for expansion and contraction of the number of regions as well as the use of alternative data sets (e.g., land cover) (see SI section 6 available at stacks.iop.org/ERL/8/015003/mmedia for potential future work with BioLUC). Expanding to additional regions would allow us to improve model resolution and precision of the results, to better evaluate regional land use dynamics, and to compare BioLUC with other LUC modeling frameworks. For example, improving the model resolution should allow for correction of unrealistic land use-related traded disparities between the US and the ROW region that appear in our results. The two-region model does not capture complex inter-regional dynamics. That is, food and biofuel demands cannot forcibly be met through imports at the expense of the internal demands of the export region, as might be expected in reality, due to different levels of purchasing power across regions. The implications of the two-region simplification are a tendency for the model to internalize land demands regionally. Modeling more regions would allow us to capture more trade complexity between developing and developed countries. For example, the general equilibrium modeling framework, GTAP, recently aggregated its data sets to 19 regions for modeling of LUC (Tyner *et al* 2010).

We selected four scenarios to represent extreme biofuel and food consumption conditions, and evaluated them to examine a high-intensity agricultural future. Many other important scenarios are possible, including:

- Additional agricultural intensification that might occur in high demand situations. That is, higher prices could lead to investment in higher yielding crops and other intensification technologies, instead of the land expansion that we found herein.
- Alternative scenarios that explore the effects of cellulosic-based biofuel production on land that is less likely to compete with food crops, such as on the abandoned land category.
- Alternative meat consumption scenarios that free up pastureland. For example, can protein and biofuel demands be met at the same time through changes in dietary preferences?

CONCLUSIONS

BioLUC results are not predictions; the model only provides insights into the dynamic interactions of LUC drivers. Quantities and changes are provided to facilitate comparisons between various scenarios.

The HF and HB scenarios lead to differing levels of cropland expansion and unequal conversion levels of pastureland and available forest and grassland requirements. In our HB scenario by 2050 cropland has expanded by 25% relative to BAU. Cropland expansion occurs mostly at the expanse of pastureland (~500 million ha), but also some available forest and grassland to compensate for used pastureland or to be used as new cropland. In our HF scenario, a 15% overall increase in cropland relative to BAU occurred, but cropland expansion onto pastureland is low relative to the HB scenario (i.e., 100 million ha) in part due to meat demand and its higher land requirements relative to other commodities. Pastureland required directly for meat production and indirectly for conversion to cropland leads to greater expansion onto available forest and grassland in the HF scenario.

The differences in land converted in the HF and HB scenarios point to alternative approximate reasons for forest and grassland conversion in the HFB scenario. In the HFB scenario the 40% increase in cropland relative to BAU is linked to biofuel and food demand in the BioLUC model. However, based on trends in the HB scenario, most of the HFB's cropland expansion is needed for biofuels. Based on our model's dynamics and relative to the HB scenario there is a stronger link between food commodities, such as meat, and the conversion of forest and grassland. Based on the comparison of HF and HB scenarios, our results suggest, all else being equal and compared to BAU, that for the HFB scenario by 2050:

- About 70% of cropland expansion is linked to higher biofuel demand.
- 30% of cropland expansion is linked to higher food demand.

The HFB scenario's high pastureland requirements led to a greater expansion into available land that is likely more directly attributed to food than to fuel so roughly 25%–35% of expansion into forest and grassland seen in the HFB scenario is attributable to biofuels. These results have potential implications for GHG emissions because forests are relatively larger carbon sinks (Hoefnagels *et al* 2010).

Our scenario analysis shows that, even under the limiting assumptions we assumed about the future, fairly aggressive future biofuel and food demands could mostly be met using a combination of agricultural and other available land. However, across all scenarios demands for the highest land-using meat commodity (e.g., beef) were difficult to meet given diet, population and other assumptions, particularly the assumption that non-meat demands (including biofuels) would be met first. Under the most extreme conditions in the HFB scenario there were some supply shortfalls relative to projected food commodity demand in the 2020–2030 timeframe. These supply shortfalls occurred when the rate of increase in food and biofuel demand was at its peak. Broad conclusions about the drivers and dynamic interactions of LUC using

BioLUC allows for an understand and test assumptions about complex systems to more informed decision making and more detailed LUC modeling analysis by other modeling systems.

Our analysis is limited in several keys respects. We did not examine additional land intensifications or advanced biofuel systems using wastes and residues that would reduce land expansion requirements. These systems could mitigate the increases in cropland area, but are not expected to alter the underlying dynamics in the current version of the model because these factors, e.g., land intensifications and using wastes and residues, could change the amount of land required but not the underlying dynamics of how land is used. One exception, which could change the underlying dynamics, is to allow cellulosic fuel feedstocks to grow on the abandoned land category, which is not currently allowed in BioLUC.

BioLUC and SD both often include general simplifications outlined in table primarily related to data and economic relationships, operating at relatively low-resolution level. Of particular importance is the lack of inter-regional dynamics in a two-region system and that price is not explicitly modeled. Model simplifications prevent the study of key dynamics emerging from or the direct result of greater model detail. With the inclusion of additional regions, trade dynamics in which greater LUC occurred in some areas over others could be explored, beyond the more proportional distribution seen in this analysis.

ACKNOWLEDGMENTS

The authors wish to acknowledge the US Department of Energy's Office of Biomass Program which provided the funding for this work. The authors do not have any other potential conflicts of interest. Data sources used in the model are included in the supplemental information, and are cited in the references section.

Thank you to Dr Robin New mark and Nate Blair of the Strategic Energy Analysis Center; Dr Joseph Cleary of the National Bioenergy Center; Dr Helena Chum a Research Fellow; Dr Doug Arent of the Institute for Strategic Energy Analysis; and Bobi Garrett Senior Vice President for Outreach, Planning, and Analysis at the National Renewable Energy Laboratory for providing internal review. Thanks to Dr Gbadebo Oladosu of Energy and Environmental Sciences at Oak

Ridge National Laboratory for external review. Special thanks also to John Sheehan at the Institute on the Environment at the University of Minnesota for reviewing the paper, for helping us define the project scope, and providing feedback on earlier drafts, and to Dana Stright for helping prepare figures for this report.

Ethan Warner made major contributions to the writing process, and also made minor contributions to data processing and analysis. Dr Daniel Inman also made major contributions to the writing process and was involved with data processing and analysis as well. Benjamin Kunstman made moderate contributions to both writing and data processing. Dr Brian Bush made major data processing and analysis contributions and made minor contributions to the paper. Laura Vimmerstedt was a minor contributor to the writing and review process. Steve Peterson was responsible for BioLUC model development and construction, and also contributed in providing project scope and analysis. Jordan Macknick provided minor data processing and writing contributions. Dr Yimin Zhang made minor writing contributions.

REFERENCES

1. Ackerman F 2002 Still dead after all these years: interpreting the failure of general equilibrium theory *J. Econ. Methodol.*9 119–39
2. Alexandratos N and Bruinsma J 2012 World agriculture towards 2030/2050: the 2012 revision *ESA Working Paper* No. 12-03 (Rome: Food and Agriculture Organization of the United Nations)
3. Al-Riffai P, Dimaranan B and Laborde L 2010 *Global Trade and Environmental Impact Study of the EU Biofuels Mandate*(Washington, DC: International Food Policy Institute)
4. Bauen A, Chudziak C, Vad K and Watson P 2010 *A Causal Descriptive Approach to Modeling the GHG Emissions Associated with the Indirect Land Use Impacts of Biofuels* (London: E4Tech)
5. Berndes G, Bird N and Cowie A 2010 Bioenergy, land use change and climate change mitigation *Bioenergy: ExCo:2010:03* (Rome: IEA Bioenergy)
6. Bush B, Inman D, Macknick J and Peterson S 2011 System dynamic approach to model land-use change from biofuels*National*

Renewable Energy Laboratory Milestone Report (Golden, CO: National Renewable Energy Laboratory)

7. California Air Resources Board (CARB) 2009 *Proposed Regulation to Implement the Low Carbon Fuel Standard*(Sacramento, CA: California Environmental Protection Agency Air Resources Board Stationary Source Division)
8. Center for BioEnergy Sustainability (CBES) 2009 *Land-Use Change and Bioenergy: Report from the 2009 Workshop*ORNL/CBES-001 (Oak Ridge, TN: Center for Bioenergy Sustainability, ORNL) (www.ornl.gov/sci/besd/cbes.shtml)
9. Chum H *et al* 2011 *Bioenergy IPCC Special Report on Renewable Energy Sources and Climate Change Mitigation* ed O Edenhofer (Cambridge: Cambridge University Press)
10. Dale V, Efroymson R and Kline K 2011 The land use–climate change–energy nexus *Landsc. Ecol.* 26 755–73
11. European Commission 2009 *Directive 2009/28/EC of the European Parliament and of the Council of 23 April 2009 On the Promotion of the Use of Energy from Renewable Sources* (Brussels: European Commission)
12. FAO 2010 *FAOSTAT Agricultural Statistics* (Rome: Food and Agriculture Organization of the United Nations)
13. Fargione J, Hill J, Tilman D, Polasky S and Hawthorne P 2008 Land clearing and the biofuel carbon debt *Science* 3191235–8
14. Farrell A E, Plevin R J, Turner B T, Jones A D, O'Hare M and Kammen D M 2006 Ethanol can contribute to energy and environmental goals *Science* 311 506–8
15. Fritsche U, Hennenberg K and Hünecke K 2010 *The iLUC factor as a Means to Hedge Risks of GHG Emissions from Indirect Land Use Change* (Darmstadt: Oeko Institute)
16. Gallagher E 2008 *The Gallagher Review of the Indirect Effects of Biofuels Production* (London: Renewable Fuels Agency)
17. Hiederer R, Ramos F, Capitani C, Koeble R, Blujdea V, Gomez O, Mulligan D and Marelli L 2010 *Biofuels: A New Methodology to Estimate GHG Emissions from Global Land Use Change—A Methodology Involving Spatial Allocation of Agricultural Land Demand and Estimation of* CO_2 *and* N_2O *Emissions* (Ispra: European Commission—Joint Research Centre)

18. Hoefnagels R E, Smeets and Faaij A 2010 Greenhouse gas footprints of different biofuel production systems *Renewable Sustainable Energy Rev.* 14 1661–94
19. ICF International 2009 *Life Cycle Greenhouse Gas Emissions Due to Increased Biofuel Production Model Linkage Peer Review Report* (Washington, DC: Environmental Protection Agency)
20. Kim H, Kim S and Dale B E 2009 Biofuels, land use change, and greenhouse gas emissions: some unexplored variables*Environ. Sci. Technol.* 43 961
21. Lobell D B and Field C B 2007 Global scale climate–crop yield relationships and the impacts of recent warming *Environ. Res. Lett.* 2 014002
22. Lywood W 2008 *Indirect Effects of Biofuels* (London: Renewable Fuels Agency)
23. NREL (National Renewable Energy Laboratory) 2012 *BioLUC Model* (Golden, CO: National Renewable Energy Laboratory)
24. Plevin R J, Michael O H, Jones A D, Torn M S and Gibbs H K 2010 Greenhouse gas emissions from biofuels' indirect land use change are uncertain but may be much greater than previously estimated *Environ. Sci. Technol.* 44 8015–21
25. Searchinger T, Heimlich R, Houghton R A, Dong F, Elobeid A, Fabiosa J, Tokgoz S, Hayes D and Yu T H 2008 Use of US croplands for biofuels increases greenhouse gases through emissions from land-use change *Science* 319 1238–40
26. Sheehan J 2009 Sustainable biofuels: a commonsense perspective on California's approach to biofuels & global land use *Indust. Biotechnol.* 5 93–103
27. Sheehan J 2012 *Reflections on the Failure of Models to Address the Land Use Impacts of Biofuels* (Saint Paul, MN: University of Minnesota Institute on the Environment)
28. Southgate D, Graham D H and Tweeten L 2007 *The World Food Economy* 1stedn(Malden, MA: Blackwell)
29. Tilman D, Balzer C, Hill J and Befort B L 2011 Global food demand and the sustainable intensification of agriculture *Proc. Natl Acad. Sci. USA* 108 20260–4

30. Tipper R, Hutchison C and Brander M 2009 *A Practical Approach for Policies to Address GHG Emissions from Indirect Land Use Change Associated with Biofuels* (Edinburgh: Ecometrica)
31. Tyner W, Taheripour F, Zhuang Q, Birur D and Baldos U 2010 *Land Use Changes and Consequent CO2 Emissions Due to US Corn Ethanol Production: A Comprehensive Analysis* (West Lafayette, IN: Purdue University, Department of Agricultural Economics)
32. UK Department for Transport 2012 *Biofuels and Sustainability Web page* (London: Department for Transport) (www.dft.gov.uk/topics/sustainable/biofuels/sustainability/)
33. US EPA (US Environmental Protection Agency) 2010 *Renewable Fuel Standard Program (RFS2) Regulatory Impact Analysis* (Washington, DC: US Environmental Protection Agency)
34. Wise M, Calvin K, Thomson A, Clarke L, Bond-Lamberty B, Sands R, Smith S J, Janetos A and Edmonds J 2009 Implications of limiting CO_2 concentrations for land use and energy *Science* 324 1183–6

Chapter 2

Consequences of Tropical Land Use for Multitrophic Biodiversity and Ecosystem Functioning

Andrew D. Barnes[1], Malte Jochum[1], Steffen Mumme[1], Noor Farikhah Haneda[2], Achmad Farajallah[3], Tri Heru Widarto[3], and Ulrich Brose[1]

[1]Department of Systemic Conservation Biology, J.F. Blumenbach Institute of Zoology and Anthropology, University of Göttingen, Berliner Strasse 28, 37073 Göttingen, Germany

[2]Department of Silviculture, Faculty of Forestry, Bogor Agricultural University, Darmaga Campus, Bogor 16680, Indonesia

[3]Department of Biology, Faculty of Mathematics and Natural Sciences, Bogor Agricultural University, Darmaga Campus, Bogor 16680, Indonesia

ABSTRACT

Our knowledge about land-use impacts on biodiversity and ecosystem functioning is mostly limited to single trophic levels, leaving us uncertain about whole-community biodiversity-ecosystem functioning relationships. We analyse consequences of the globally important land-use transformation from tropical forests to oil palm plantations. Species diversity, density and biomass of invertebrate communities suffer at least 45% decreases from rainforest to oil palm. Combining metabolic and food-web theory, we calculate annual energy fluxes to model impacts of land-use intensification on multitrophic ecosystem functioning. We demonstrate a 51% reduction in energy fluxes from forest to oil palm communities. Species loss clearly explains variation in energy fluxes; however, this relationship depends on land-use systems and functional feeding guilds, whereby predators are the most heavily affected. Biodiversity decline from forest to oil palm is thus accompanied by even stronger reductions in functionality, threatening to severely limit the functional resilience of communities to cope with future global changes.

INTRODUCTION

The transformation from natural ecosystems to agricultural land use and its continued intensification has led to extensive losses in biodiversity and ecosystem services[1] resulting in the degradation of human well being[2]. The transformation of lowland tropical rainforest to oil palm (*Elaeis guineensis* Jacq.) plantations has gained more recent attention as an especially severe threat to tropical biodiversity[3, 4]. In the last 25 years the total plantation area of oil palm has tripled, with current global estimates of over 15 million hectares[3], making this crop one of the world's most rapidly expanding forms of agriculture[5]. It is now clear that the expansion of oil palm agriculture is one of the greatest causes of deforestation[6, 7], and this threat appears to be increasing without respite as Indonesia, one of the world's leaders in oil palm, makes plans to double production by 2020 (ref. 8). The rapid expansion of such large-scale land-use transformation raises questions about the impending implications for biodiversity and ecosystem functioning in the tropics.

Despite a broad consensus that biodiversity is positively correlated with ecosystem functioning in controlled experiments[9, 10], there are few real-world examples of such biodiversity–ecosystem functioning relationships[11,12]. In fact, until now there have been no studies that explore the relationship between biodiversity and ecosystem functioning in ecosystems undergoing agricultural land-use transformation to oil palm. Thus, our knowledge of this globally important land-use conversion is strongly limited. Furthermore, over the past decade there have been important advances towards multitrophic approaches in research investigating biodiversity–ecosystem functioning relationships[9, 13, 14, 15, 16]. Despite these advances, however, we are still substantially limited by the lack of clear approaches to quantify single measures of ecosystem functioning that can be compared among any combination of trophic levels. This has resulted in our inability to directly look at whole-community relationships between entire species assemblages and the respective functional processes carried out in these communities.

Here we use the total energy flux between functional feeding guilds as a measure of multitrophic ecosystem functioning, as many studies have suggested process rates, such as energy fluxes, to be important proxies for ecosystem functioning[10, 13, 17]. Depending on the resource pool that the energy flux comes from, these fluxes can be directly related to ecosystem services such as decomposition[18, 19], plant biomass production[20, 21] or biocontrol through predation[22]. These energy flux calculations are based on metabolic scaling theory[23] and principles of food-web energy dynamics[18]. Using individual metabolic rates that are dependent on body mass, environmental temperature and phylogenetic grouping[18, 24], combined with resource-specific assimilation efficiencies[25] and energy loss to predation[18], we present this energy flux calculation as a unified measure of multitrophic ecosystem functioning (Fig. 1). Studies that incorporate diversity across trophic levels to test the relationship between biodiversity and ecosystem functioning have predominantly used only biomass as the measure of ecosystem function[26]. However, the metabolic activity and thus the energy-processing rates of these biomass pools can vary substantially. Integrating over body mass, phylogeny and temperature with their constraints on metabolic rates, and additionally taking into account assimilation efficiencies and loss to predation, our measure of whole-community energy flux inherently incorporates not only biomass but also other important ecosystem attributes enabling the quantification of emergent functional properties of ecosystems that would otherwise remain undetected. As such, our measure of energy flux provides a comprehensive and robust measure of multitrophic ecosystem functioning that can be utilized for modelling biodiversity–ecosystem functioning relationships for any assemblage of taxonomic groups, while incorporating multiple ecological functions.

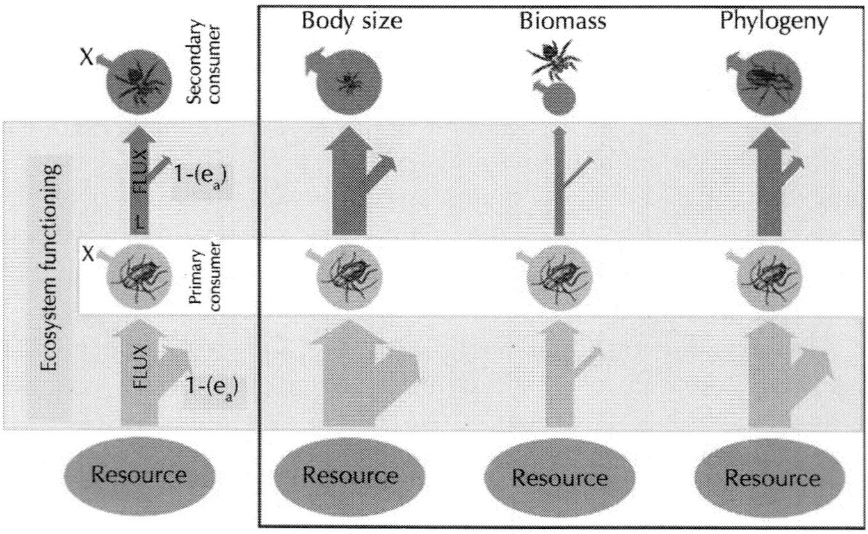

Figure 1: Energy fluxes along a conceptual food chain as a measure of multi-trophic ecosystem functioning.

Energy flux between two nodes is calculated as $F = \frac{1}{e_a} \cdot (X + L)$, where *F* is the total energy flux into the network node of a feeding guild (vertical red and yellow arrows), e_a is the diet-specific assimilation efficiency (denoted by diagonal arrows arising from the flux arrows), *X* is the per-unit-mass metabolic demand of the feeding guild (which is nonlinearly dependent on body sizes, temperature and phylogeny) and *L* is the loss to predation from the node (for the yellow node, this is equal to the flux to the red secondary consumer node). Here we demonstrate three examples where changes in the mean body size (size of black animal icons), biomass (diameter of red and yellow circles) or phylogeny (black animal icons) on any trophic level (here demonstrated by the secondary consumer guild) can result in nonproportionally altered total energy flux (sum of all arrow widths in the food chain).

In the tropical lowland rainforests of Sumatra, Indonesia, which have been undergoing vast land-use transformation to oil palm[7], we quantify the impacts of this transformation ranging from tropical secondary rainforest, jungle rubber and intensively managed rubber, to oil palm. We utilize data gathered from 32 sites in Sumatra, Indonesia, comprising 2,415 populations of 871 species. First, we investigate the biodiversity value of jungle rubber, conventional rubber and secondary forest compared with oil palm agriculture by comparing observed species richness, density and biomass of litter-associated macroinvertebrate communities across these systems. Second, as a multitrophic measure of the rate of ecosystem processes carried out by these communities, we calculate total solid fresh mass energy flux in a system by incorporating community metabolism[27], resource-specific assimilation efficiencies and biomass loss to predation[18] into whole-community energy flux equations (Fig. 1). This provides a quantitative measure of multitrophic ecosystem functioning, defined here as the total flux of energy from any resource pool to consumer trophic levels. In addition, this measure can be attributed to specific functional feeding guilds within communities to look for patterns in ecosystem functioning at different trophic levels. Using the energy-mass flow conversion[28], we express energy flux as kilograms per hectare, per year and explore the relationship between total species diversity and energy flux, distinguishing among four transformation systems to test for land-use-dependent biodiversity–ecosystem functioning relationships. Our results demonstrate strong losses in species diversity that, in turn, predict reductions in whole-community energy fluxes. However, these reductions are strongest in oil palm systems, suggesting that land-use conversion from forest to oil palm causes disproportionally strong losses in multitrophic ecosystem functioning.

RESULTS

Transformation to Oil Palm Leads to Biodiversity Loss

Using generalized linear mixed effects models, we show that transformation of tropical rainforest to oil palm plantations leads to severe losses in species richness (45% decline), animal density (48% decline) and biomass (52% decline; Fig. 2a–c and Supplementary Table 1), supporting previous studies suggesting that land-use transformation to oil palm poses one of the greatest threats to global biodiversity[3]. Beyond mere diversity effects, land-use transformation altered animal densities and biomass, threatening to not only drive species extinctions but also to eliminate vital ecological functions. The effects of land-use transformation on species richness and animal densities were additionally dependent on functional feeding guilds, with predators decreasing in species richness and density most rapidly (Fig. 2a–c and Supplementary Table 1) as could be expected for higher trophic level feeding guilds[29]. Such alteration of higher trophic levels is likely to have severe indirect functional impacts on other functional guilds within the trophic network[30].

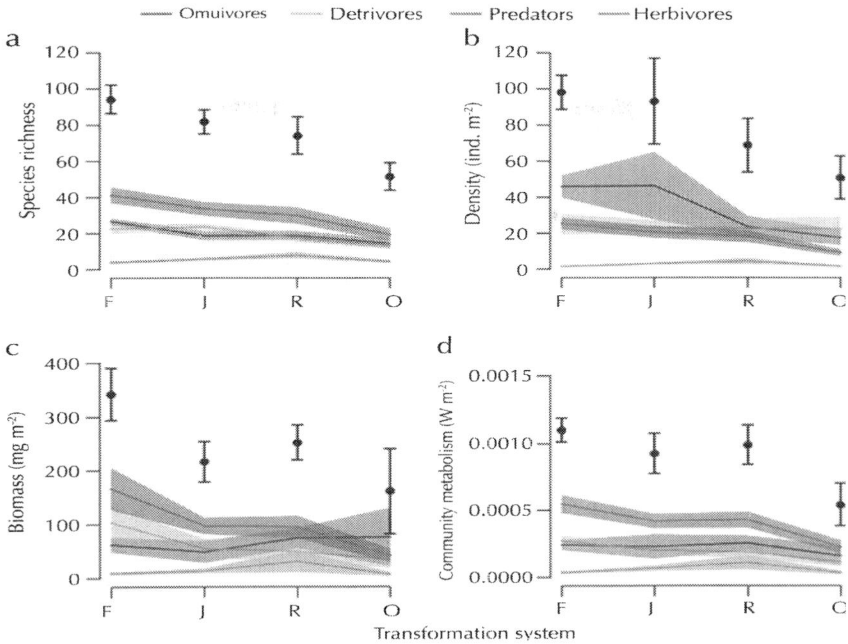

Figure 2: Effects of land-use transformation on macroinvertebrate communities. The mean (±s.e., *n*=32) species richness (a), density (b), biomass (c) and community metabolism (d) of the total community (black points) and of each functional feeding guild (coloured lines) for the four land-use transformation systems: forest (F), jungle rubber (J), rubber (R) and oil palm (O).

Community Metabolism

Summing up individual metabolic rates, we demonstrate that transformation of forest to oil palm yields a 51% decrease in community metabolism, with jungle rubber and rubber only 16% and 10% below forest levels of community metabolism, respectively. However, all systems yielded significantly higher community metabolism than oil palm (Fig. 2d and Supplementary Table 1). As such, we show that ecosystem energy processing is critically reduced in oil palm plantations. Interestingly, biomass responses to land-use transformation among feeding guilds were not clearly comparable to responses in community metabolism (Fig. 2c,d). This suggests that systematic changes in species composition, body-mass distributions

(Supplementary Fig. 1) and biomass exhibited a complex interaction in determining the functional consequences of land-use transformation.

Whole-Community Energy Fluxes and Ecosystem Functioning

Aiming to visualize the complex interplay between community biomass dynamics and energy flux, we constructed energy networks for the four transformation systems (Fig. 3) based on total energy fluxes as a promising way to quantify multitrophic ecosystem functioning (Fig. 1). In addition to the general decreases in biomass (node sizes in Fig. 3) and energy-processing rates (arrow widths in Fig. 3), we also found a systematic shift from predator to omnivore dominance when comparing forest and oil palm systems. Specifically, we found that predator biomass in oil palm yielded only 25% of their biomass in forest (0.424 and 1.664 kg ha^{-1}, respectively), while the predator-driven energy flux was reduced to 46% of the energy flux driven by predators in forest (30.697 and 66.816 kg ha^{-1} per year, respectively). In contrast, omnivore biomass in oil palm was 22% higher than in the forest (0.767 compared with 0.629 kg ha^{-1}), while omnivore-driven energy flux in the oil palm was 47% lower than in forest communities (32.531 compared with 61.900 kg ha^{-1} per year; Supplementary Table 2), suggesting a considerable mismatch of biomass and energy flux, partly dependent on the trophic group in question. In our analyses, this disparity finds its explanation in varying body-mass distributions (Supplementary Fig. 1) and assimilation efficiencies that strongly modify how biomass translates into total resource assimilation rates (Fig. 1). These results suggest that biomass, alone, may be an unsuitable proxy for general ecosystem functioning in animal communities.

Figure 3: Effects of land-use transformation on community energy networks. Energy networks displaying the relative annual energy flux (coloured arrow width weighted by calculated energy flux (kg ha^{-1} per year)) and biomass (coloured node diameter weighted by total biomass) among the functional feeding guilds: predators (red), omnivores (blue), detritivores (yellow) and herbivores (green). Each panel represents an energy network for one of the four land-use transformation systems.

Multitrophic Biodiversity-Ecosystem Function Relationships

Until now, most studies investigating biodiversity–ecosystem function relationships have focused on single trophic levels[31, 32]. We present a new approach to easily quantify multitrophic ecosystem functioning, requiring only information on body mass, phylogeny, temperature and assimilation efficiencies to overcome previous limitations in biodiversity–ecosystem functioning research. Utilizing this approach, we also investigated the relationship between species richness and ecosystem functioning, identifying a clear linear positive effect of diversity on total energy flux (Fig. 4a and Supplementary Table 3). The relationship between diversity and energy flux was dependent on land-use transformation system, whereby oil palm and jungle rubber showed the strongest decrease in energy flux per unit loss in species richness

(Fig. 4a and Supplementary Table 3). Our results suggest that each loss of species in oil palm and jungle rubber therefore would be followed by proportionately higher losses in energy flux, compared with equal species losses in forest and rubber. We found the same pattern as in the overall trend for the predator group, which showed transformation system-dependent relationships between species richness and energy flux (Fig. 4b). However, for omnivores, detritivores and herbivores there was a linear effect of diversity on energy flux driven by these groups; however, this effect was independent of transformation system (Fig. 4b and Supplementary Table 3). This implies that studies focusing on single trophic levels, or even specific species, may fail to detect the alteration of ecosystem processes resulting from land-use transformation. These results call for a wider application of multitrophic approaches that not only measure one ecosystem property, such as total productivity or decomposition, but that also aim to assess whole-community ecosystem processes such as total energy flux.

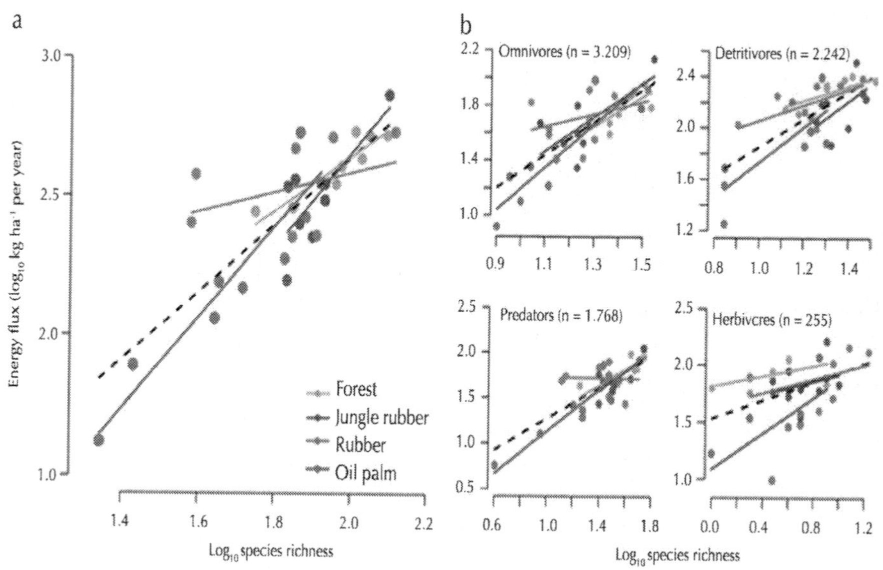

Figure 4: Relationship between species richness and community energy fluxes. Linear mixed effects models for (a) entire communities and (b) separated into functional feeding guilds. Black dashed lines denote overall model fits and coloured lines indicate different land-use transformation systems.

DISCUSSION

Our study reflects previous findings that the transformation of forest systems to oil palm has severe impacts not only on single animal populations but also on communities as a whole. In particular, species richness and animal biomass are most significantly affected. Furthermore, jungle rubber and rubber appear to represent intermediate steps in land-use intensification. Their higher levels of biodiversity and ecosystem functioning indicate that they potentially provide higher ecological value than oil palm. As such, these rubber land-use systems could present economically viable, lower intensity land-use alternatives.

By taking a multitrophic ecosystem functioning approach we demonstrate that, at the community level, species loss leads to a direct linear decrease in ecosystem functioning. This means that any species loss will be followed by a proportionate loss in function, and this relationship becomes proportionately stronger in more intensive transformation systems such as oil palm plantations. Thus, every one of the few species in high-intensity land-use systems is functionally more important than species in low-intensity systems where functional redundancy is likely to be higher[33]. Without explicit consideration of multiple trophic levels, such emergent properties are likely to be overlooked. Our study demonstrates the crucial implications of tropical land-use intensification for biodiversity and ecosystem functioning across multiple trophic levels, suggesting that these globally important impacts will likely resonate beyond previously explored trophic boundaries.

METHODS

Study Site and Sampling Design

Sampling took place in the Jambi province of Sumatra, Indonesia, a region known as a hotspot for biodiversity, but that has also already undergone extensive deforestation[6, 34]. In the second half of the last century, Sumatra's forests have experienced vast transformation to rubber and oil palm monocultures[35, 36]. This large-scale land-use conversion has left Sumatra with a very limited area of natural forest mainly restricted to national parks and even here, where logging has been reduced, it has not come to a complete halt[37]. This severe and extensive land-use transformation, that has progressed already further than in most other tropical landscapes, makes Sumatra a unique and ideal example system for studying the impacts of land-use conversion on biodiversity and ecosystem functioning.

We sampled secondary rainforest, jungle rubber, rubber and oil palm systems, replicated eight times across two landscapes (n=32; Supplementary Fig. 2). Sites were selected by first looking for landscapes in the Jambi province that still contained secondary rainforest. Second, we identified all lowland areas with little or no slope and then randomly selected two landscapes with 16 sites each. Among all of the 32 sampling sites, we maintained a minimum distance of 120 m to insure independence of the epigaeic invertebrate communities sampled. The secondary-forest regions lie within two protected areas, Bukit Duabelas National Park and Harapan Rainforest, and represent the least influenced land-use system. Jungle rubber—forest stands with a high percentage of rubber trees that are still regularly harvested—represents a low-impact agroforestry system[38]. Rubber and oil palm plantations serve as locally common[36] high-impact monocultures. The 32 sites were carefully selected so that they were all of a similar age and from equal elevations close to the sea level. All agricultural systems (jungle rubber, rubber and oil palm) were treated and harvested by their owners with intensities typical for the respective transformation system.

Animal Sampling and Calculation of Response Variables

Animal sampling took place between early October and early November 2012. All organisms were collected based on Permit No. 51/KKH-5/TRP/2014 issued by the Indonesian Institute of Sciences and the Ministry of Forestry. In all 32 of the 50 × 50-m sites, we sampled once in each of three 5 × 5-m subplots by sieving the leaf litter from 1 m^2 through a coarse sieve of 2-cm-width mesh. In all, 7,472 macroinvertebrates were hand-collected from the sieving samples and stored in 65% ethanol. Specimens were identified to morphospecies and assigned to one of four feeding guilds: omnivores, detritivores, predators and herbivores, based on morphology and literature.

As biodiversity studies always suffer from undersampling and correlation of sample size with species richness, we compared observed species richness to both extrapolated and rarefied species richness, calculated in the 'vegan' package in R[39], to assess the accuracy of our species-sampling effort. To extrapolate sampled species richness, we used the nonparametric second-order jacknife estimator[40] to calculate extrapolated species richness from the three 1-m^2subsamples at each of the 32 sites, revealing an estimated mean sampling coverage of 56% (s.d. of±2.393%) making the second-order jacknife estimator the most accurate extrapolation method[40]. In addition, we calculated sample-based rarefaction, whereby rarefaction curves were calculated for each of the 32 sampled sites and then cut off at the sample size of the smallest sample (40 individuals). Because of the very high attrition of data during the rarefaction procedure (a total of 6,192 out of 7,472 individuals, or 83%, were removed), the rarefied species richness yielded very little resemblance to observed species richness when comparing across transformation systems, resulting in almost no pattern of rarefied richness among transformation systems (Supplementary Fig. 3). The jacknife2-extrapolated species richness, however, was extremely closely correlated with observed species richness (Pearson's ρ=0.993) patterns among transformation systems (Supplementary Fig. 3), suggesting that our observed species richness did in fact accurately capture realistic patterns in total species diversity across the land-use transformation systems.

For each of the 7,472 animals collected, we measured individual body length to an accuracy of 0.1 mm using stage micrometres. We then converted all measured individual body lengths to fresh body mass using length-mass regressions and, where necessary, dry mass-fresh mass relationships from the literature (Supplementary Table 4), yielding an estimated fresh mass in milligrams for every collected individual. Where family-specific relationships were not available or animal body lengths in our collection fell outside of the size ranges of published regressions, we then used regressions from higher-order taxonomic groupings. For heavily damaged individuals that could not be measured for body length, we assigned these individuals a fresh body mass from the median body mass of all animals from the same species or order where only one individual of that species was collected. We then calculated community biomass (mg fresh mass m^{-2}) for each of the 32 communities by summing together all individual body masses calculated from length-mass regressions as derived from the individually measured body lengths.

We calculated individual metabolic rates for all 7,472 animals using body masses, temperature and phylogeny[24] (Supplementary Table 5). Temperature was measured over a period of at least 2.5 months at 30-cm depth below the soil surface in each site and averaged for each transformation system in each of the two landscapes. From this, community metabolism was calculated by summing together all individual metabolic rates within each of the 32 sites, providing the total metabolic demand for each of the 32 communities. Using diet-specific assimilation efficiencies[25], energy loss to predation and community metabolism, we analytically calculated energy fluxes for each of these communities[18] using the formula

$$F = \frac{1}{e_a} \cdot (X + L) \qquad (1)$$

where *F* is the total energy flux into the network node of a feeding guild, e_a is the diet-specific assimilation efficiency, *X* is the metabolic demand of the feeding guild and *L* is the loss to predation that the feeding guild is subjected to (Fig. 1 and Supplementary Methods). In order to calculate the fluxes between the functional feeding guilds, we constructed a general network of feeding relationships (link structure in Fig. 3) that represents a null model for an energy network structure where no active preferences are assumed. We assumed that, of our four functional feeding guilds, energy fluxes to predators were split up equally into the three animal guilds below them. Energy fluxes to detritivores and herbivores were assumed to come from only detritus and plant material, respectively. Omnivores were assumed to receive energy in equal 25% proportions from the other three functional feeding groups (predators, detritivores and herbivores, making 75%) and the remaining 25% from both plant and detritus material combined (Supplementary Methods).

To assess how these assumptions of feeding preferences might affect the calculations of total energy fluxes, we reconstructed the energy networks so that omnivores were assumed to only consume plant and detritus material (50% derived from each) but with no energy derived from animal material. We then recalculated total energy fluxes and found an overall decrease of up to 54%, which appeared to be highly consistent among the different land-use transformation systems. This consistency between models was especially evident after calculating the loss of energy flux in the three agriculturally used systems compared with the forest system, demonstrating a maximum of only 3% disparity between the two models (Supplementary Fig. 4). This sensitivity analysis indicated that our presented method is highly robust in calculating differences in energy fluxes among different systems. Accordingly, the null model was accepted as the simplest model with the least diet preferences assumed. However, we still suggest that studies adopting this method of energy flux calculation should assign feeding preferences with caution, or employ other techniques such as stable isotope analysis to estimate feeding preferences.

Statistical Analyses

Using mixed effects models (GLMM's), we tested the effects of 'transformation system' and its interaction with functional feeding guild on community responses, with 'landscape' as a random effect. 'Density', 'biomass' and 'community metabolism' were $\log_{10}$-transformed to meet assumptions of normality and 'species richness' (overdispersed poisson-distributed data) was modelled on a negative binomial distribution. We additionally explored biodiversity–ecosystem functioning relationships by first testing for linearity of relationships using untransformed data. Once linearity was established, we then tested for the effects of $\log_{10}$-transformed 'species richness' and its interaction with 'transformation system' on 'energy flux' for overall data and repeated again for data from separate feeding guilds. In addition, because we suspected that our analyses could be affected by spatial autocorrelation, we calculated Moran's *I* values for each model's residuals and tested for spatial autocorrelation using the Moran's *I* standard deviate[41] in the 'spdep' package in R 3.0.2 (ref. 39). Results from these tests provided no support for the spatial autocorrelation of variation in any of the response variables tested (all Moran's *I* test results yielded $P>0.4$).

For all GLMM's, we applied a backward stepwise selection procedure to obtain the model of best fit, based on the Akaike

Information Criterion (AIC). In this procedure, we constructed full models that contained all possible predictors and their interactions ('transformation system' and 'feeding guild' for general community response models; 'species richness' and 'transformation system' for biodiversity–ecosystem functioning models) and compared these full models and the model of the backward selection procedure to a null, intercept-only model. The model that yielded the lowest AIC score, with a minimum ΔAIC of 2 units, was selected as the model of best fit. All analyses were conducted with the 'nlme' and 'lme4' packages in R 3.0.2 (ref. 39).

ACKNOWLEDGEMENTS

We thank Megawati, Rizky Nazarreta, Keisha Disa Putirama, Rosario Reza Valentino Lasse for assistance in the field and laboratory. Roswitha Ehnes provided additional metabolic rate regression parameters and Christian Guill assisted in solving energy flux equations. Ana Meijide, Alexander Knohl, Oleg Panferov and team provided climate data. We also thank the village leaders, local site owners, PT REKI and Bukit Duabelas National Park for granting us access to their properties. This study was financed by the Deutsche Forschungsgemeinschaft (DFG) in the framework of the collaborative German—Indonesian research project CRC990.

REFERENCES

1. Gibbs, H. K. *et al.* Tropical forests were the primary sources of new agricultural land in the 1980s and 1990s. *Proc. Natl Acad. Sci. USA* 107, 16732–16737 (2010).
2. Díaz, S., Fargione, J., Chapin, F. S. & Tilman, D. Biodiversity loss threatens human well-being. *PLoS Biol.* 4, e277 (2006).
3. Gilbert, N. Palm-oil boom raises conservation concerns. *Nature* 487, 14–15 (2012).
4. Koh, L. P. & Wilcove, D. S. Cashing in palm oil for conservation. *Nature* 448, 993–994(2007).
5. Fitzherbert, E. B. *et al.* How will oil palm expansion affect biodiversity? *Trends Ecol. Evol.* 23,538–545 (2008).

6. Wilcove, D. S., Giam, X., Edwards, D. P., Fisher, B. & Koh, L. P. Navjot's nightmare revisited: logging, agriculture, and biodiversity in Southeast Asia. *Trends Ecol. Evol.* 28,531–540 (2013).
7. Koh, L. P., Miettinen, J., Liew, S. C. & Ghazoul, J. Remotely sensed evidence of tropical peatland conversion to oil palm. *Proc. Natl Acad. Sci. USA* 108, 5127–5132 (2011).
8. Koh, L. P. & Ghazoul, J. Spatially explicit scenario analysis for reconciling agricultural expansion, forest protection, and carbon conservation in Indonesia. *Proc. Natl Acad. Sci. USA* 107, 11140–11144 (2010).
9. Cardinale, B. J. *et al.* Effects of biodiversity on the functioning of trophic groups and ecosystems. *Nature* 443, 989–992 (2006).
10. Hooper, D. U. *et al.* Effects of biodiversity on ecosystem functioning: a consensus of current knowledge. *Ecol. Monogr.* 75, 3–35 (2005).
11. sFoster, W. A. *et al.* Establishing the evidence base for maintaining biodiversity and ecosystem function in the oil palm landscapes of South East Asia. *Philos. Trans. R. Soc. Lond. B. Biol. Sci.* 366, 3277–3291 (2011).
12. Otto, S. B., Berlow, E. L., Rank, N. E., Smiley, J. & Brose, U. Predator diversity and identity drive interaction strength and trophic cascades in a food web. *Ecology* 89, 134–144 (2008).
13. Duffy, J. E. Biodiversity and ecosystem function: the consumer connection. *Oikos* 99,201–219 (2002).
14. Petchey, O. L. *et al.* Species loss and the structure and functioning of multitrophic aquatic systems. *Oikos* 104, 467–478 (2004).
15. Schneider, F. D., Scheu, S. & Brose, U. Body mass constraints on feeding rates determine the consequences of predator loss. *Ecol. Lett.* 15, 436–443 (2012).
16. Schneider, F. D. & Brose, U. Beyond diversity: how nested predator effects control ecosystem functions. *J. Anim. Ecol.* 82, 64–71 (2013).
17. Srivastava, D. S. & Vellend, M. Biodiversity-ecosystem function research: Is it relevant to conservation? *Annu. Rev. Ecol. Evol. Syst.* 36, 267–294 (2005).
18. De Ruiter, P. C., Neutel, A. M. & Moore, J. C. Modelling food webs and nutrient cycling in agro-ecosystems. *Trends Ecol. Evol.* 9, 378–383 (1994).

19. Handa, I. T. *et al.* Consequences of biodiversity loss for litter decomposition across biomes.*Nature* 509, 218–221 (2014).
20. Enquist, B. J. *et al.* A general integrative model for scaling plant growth, carbon flux, and functional trait spectra. *Nature* 449, 218–222 (2007).
21. Tilman, D., Reich, P. B. & Knops, J. M. H. Biodiversity and ecosystem stability in a decade-long grassland experiment. *Nature* 441, 629–632 (2006).
22. Cardinale, B. J., Harvey, C. T., Gross, K. & Ives, A. R. Biodiversity and biocontrol: emergent impacts of a multi-enemy assemblage on pest suppression and crop yield in an agroecosystem. *Ecol. Lett.* 6, 857–865 (2003).
23. Brown, J. H., Gillooly, J. F., Allen, A. P., Savage, V. M. & West, G. B. Toward a metabolic theory of ecology. *Ecology* 85, 1771–1789 (2004).
24. Ehnes, R., Rall, B. & Brose, U. Phylogenetic grouping, curvature and metabolic scaling in terrestrial invertebrates. *Ecol. Lett.* 14, 993–1000 (2011).
25. De Ruiter, P. C., Veen, J. a., Moore, J. C., Brussaard, L. & Hunt, H. W. Calculation of nitrogen mineralization in soil food webs. *Plant Soil* 157, 263–273 (1993).
26. Duffy, J. E. *et al.* The functional role of biodiversity in ecosystems: incorporating trophic complexity. *Ecol. Lett.* 10, 522–538 (2007).
27. Ehnes, R. *et al.* Lack of energetic equivalence in forest soil invertebrates. *Ecology* 95,527–537 (2014).
28. Peters, R. H. *The Ecological Implications of Body Size* Cambridge Studies in Ecology329(1983).
29. Purvis, A., Gittleman, J. L., Cowlishaw, G. & Mace, G. M. Predicting extinction risk in declining species. *Proc. Biol. Sci.* 267, 1947–1952 (2000).
30. Jochum, M., Schneider, F. D., Crowe, T. P., Brose, U. & O'Gorman, E. J. Climate-induced changes in bottom-up and top-down processes independently alter a marine ecosystem.*Philos. Trans. R. Soc. Lond. B. Biol. Sci.* 367, 2962–2970 (2012).
31. Balvanera, P. *et al.* Quantifying the evidence for biodiversity effects on ecosystem functioning and services. *Ecol. Lett.* 9, 1146–1156 (2006).

32. Ives, A. R., Cardinale, B. J. & Snyder, W. E. A synthesis of subdisciplines: predator-prey interactions, and biodiversity and ecosystem functioning. *Ecol. Lett.* 8, 102–116 (2004).
33. Laliberte, E. *et al.* Land-use intensification reduces functional redundancy and response diversity in plant communities. *Ecol. Lett.* 13, 76–86 (2010).
34. Sodhi, N. S., Koh, L. P., Brook, B. W. & Ng, P. K. L. Southeast Asian biodiversity: an impending disaster. *Trends Ecol. Evol.* 19, 654–660 (2004).
35. Wilcove, D. S. & Koh, L. P. Addressing the threats to biodiversity from oil-palm agriculture.*Biodivers. Conserv.* 19, 999–1007 (2010).
36. Laumonier, Y. *et al.* Eco-floristic sectors and deforestation threats in Sumatra: identifying new conservation area network priorities for ecosystem-based land use planning. *Biodivers. Conserv.* 19, 1153–1174 (2010).
37. Gaveau, D., Wandono, H. & Setiabudi, F. Three decades of deforestation in southwest Sumatra: have protected areas halted forest loss and logging, and promoted re-growth?*Biol. Conserv.* 134, 495–504 (2007).
38. Gouyon, A., Foresta, H. & Levang, P. Does jungle rubber deserve its name? An analysis of rubber agroforestry systems in southeast Sumatra. *Agrofor. Syst.* 22, 181–206 (1993).
39. R Development Core Team. R: A language and environment for statistical computing. *R Found. Stat. Comput.* (2014) available at http://www.r-project.org/.
40. Brose, U., Martinez, N. D. & Williams, R. J. Estimating species richness: sensitivity to sample coverage and insensitivity to spatial patterns. *Ecology* 84, 2364–2377 (2003).
41. Dormann, C. F. *et al.* Methods to account for spatial autocorrelation in the analysis of species distributional data: a review. *Ecography* 30, 609–628 (2007).

Chapter 3

A Comparison of Two Land Use Simulation Models under the RCP4.5 Scenario in China

Feng Wu[1], Jinyan Zhan[1], Qian Xu[2], Yihui Xiong[3], and Zhongxiao Sun[2]

[1]State Key Laboratory of Water Environment Simulation, School of Environment, Beijing Normal University, Beijing 100875, China

[2]School of Mathematics and Physics, China University of Geosciences, Wuhan 430074, China

[3]School of Information Engineering, China University of Geosciences, Wuhan 430074, China

ABSTRACT

The land use simulation model is an important tool to analyze the land use/land cover change (LUCC), which plays a key role in influencing the global warming. However, there have been very few global LUCC simulation models, especially the models that can be used to analyze the interaction among the socioeconomic development, climate change, and LUCC. The Global Change Assessment Model (GCAM) and the GTAP-AEZ model are two models that take account of the influence of social economy and climate change at the global scale, but they may have some parameter errors due to the rough parameter setting. This study aims to compare the simulation results obtained with the GCAM model and GTAP-AEZ model and optimize their parameters according to the specific conditions of China. First, we simulated the land use structure in China in 2010 with the two models and compared the simulation results with the real one. Second, we calibrated these parameters of models according to the China's national conditions and implemented the simulation again. The result indicates that the calibrated GCAM can provide more accurate simulation result of land use, which can provide significant reference information for the land use planning and policy formulation to mitigate the climate change in China.

INTRODUCTION

Humans have transformed significant portions of the Earth's land surface, 10 to 15 percent of which is currently dominated by agricultural crop or urban-industrial areas, and 6 to 8 percent is pasture [1]. These land use changes have important implications for future climate changes and, consequently, great implications for subsequent land use changes [2–4]. Climate change and land use change are both global drivers of environmental change, and the impact assessments generally show that interactions between them can lead to serious challenges to the provision of ecosystem services. Besides, in many cases it is impossible to determine the impacts of climate change without consideration of land use/land cover change (LUCC). LUCC is a widespread, accelerating, and significant process, and it has been one of the research cores of the international programmes such as the International Geosphere-Biosphere Programme (IGBP) and the Global Environmental Change Human Dimensions Programme (IHDP) and is also one of the global environmental research focuses and cutting-edge issues [5]. LUCC is driven by human activities, and in many cases it also leads to changes that impact the humans; therefore, LUCC modeling is a critical way for formulating effective environmental policies and management strategies [6, 7]. Understanding the role of land use change in the global environmental change requires the analysis of historical land cover conversions and/or projection of possible future land use changes, both of which heavily rely on the land use simulation models. Besides, the land use simulation model also provides an essential approach for stakeholder to project and evaluate the potential consequences of policy decisions and other actions. As more scholars realized the importance of LUCC, the land use simulation model has become an important tool for the analysis of both the mechanism and the spatial distribution of LUCC in the past and future [8, 9]. The land use simulation models include Markov model, logistic function model, regression model, econometric model, dynamic systems model, spatial simulation model, linear planning model, nonlinear mathematical planning model, mechanistic GIS model, and cellular automata model [10]. All of these models may help to explore the combined effects of social policies, individual behavior, and other drivers of the land use change; however, most of them have some drawbacks. For example, the Markov model has been widely used to simulate the land use change, but it involves no spatial factors, and the land use change cannot be spatially and explicitly reflected. The CLUE-S model can comprehensively analyze the regional LUCC process and driving force, but it can only be used in the spatial allocation of land use changes so far, while the nonspatial changes must be estimated with other methods [11]. Therefore, although some models can be used to simulate land use change, there are still some serious drawbacks [12, 13]. Moreover, there were few global models to simulate the LUCC, especially in the study of the interaction mechanism among the social economy, climate change, and LUCC. In some sense, the GTAP-AEZ model and the Global Change Assessment Model (GCAM) are more useful in the land use simulation, which can simulate the land use change of each agricultural ecological zone (AEZ), combined with the influences of social economy and climate change at global scale [14–16]. However, the parameters of these models are rough, and the simulation accuracy needs to be improved.

In the preparation for the fifth assessment report of the Intergovernmental Panel on Climate Change (IPCC AR5), the international community is developing new advanced Earth System

Models (ESMs) to address the combined effects of human activities (e.g., land use and greenhouse gas emissions) on the carbon-climate system. Besides, the four representative concentration pathways (RCPs) scenarios of the future (2005–2100) have been provided by the four Integrated Assessment Model (IAM) teams, which are used as input to the ESMs for the future carbon-climate projection [17, 18]. This study aims to compare the simulation results of land use change obtained with the GCAM and GTAP-AEZ models and improve the simulation accuracy through optimizing the input parameters of the models, and the calibrated GCAM can be used to provide more accurate reference information of land use change for the land use planning and policy formulation to mitigate the climate change in China.

MODELS

GCAM

GCAM is a dynamic recursive model of land use and land cover, economy, agriculture, and energy, which fully integrates the energy and agriculture systems with economic equilibrium in the energy and agriculture markets [19]. GCAM consists of four modules, that is, Edmonds-Reilly-Barnes (ERB) model [20, 21], Agriculture and Land Use (AgLU) simulation model [22,23], Model for the Assessment of Greenhouse-gas Induced Climate Change (MAGICC) [24], and Regional Climate Change Scenario Generator (SCENGEN) [25]. The inputs of GCAM include capital, labor, and initial land use allocation, all of which need to be provided by researchers.

The land use allocation framework shows how the land area is allocated among alternative land use types (Figure 1). The markets in GCAM are defined for energy and agricultural products, which are driven by factors such as the population, labor productivity growth, and price of resources. The land use allocation is achieved through the price mechanism in the market (Figure 1); that is, the land use allocation is based on maximization of the economic return in a region. The economic return is represented with the profit per hectare, which is the revenue (yield per hectare times price received) minus the production cost (yield per hectare times nonland cost per unit of input), as is shown in

$$\pi r_{i,l,m,p} = y_{i,l,m,p} \bullet (P_{i,l,m,p} - G_{i,l,m}), i, p, l \quad (1)$$

where $\pi r_{i,l,m,p}$ is the economic return of the land measured with the profit rate ($/ha·yr), $y_{i,l,m,p}$ is yield per hectare of the land use type i in the AEZ P (calories/ha), $P_{i,l,m}$ is the market price of the product produced by the land use type , is the nonland cost per unit output, i is an index for land use type, l $G_{i,l,m}$ is the region index, and P is an index for agricultural ecological zone (AEZ) within a region.

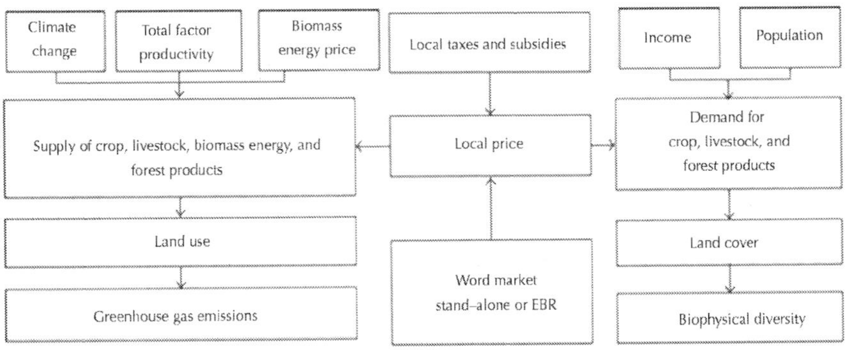

Figure 1: Land use allocation framework of GCAM.

The calculation of the profit rate (πr) of forest products is somewhat different due to the time lag between planting and harvest, and it includes a term that converts the future earnings into the present earnings as follows:

$$\pi r_{i,l,m,p} = \frac{r}{(1+r)^{45} - 1} \bullet (\bar{P}_{i,l,m} - G_{i,l,m}), \quad (2)$$

Where r is the interest rate ($/$, that is dimensionless).

In order to determine the land area allocated to each land use type, the shares of land use were numerically analyzed based on the profit rates calculated with (1) and (2). Then a simplified approach was used to allocate the land area in each AEZ with the aim of maximizing the profit, with which the maximum profit rate after the land use allocation in each AEZ can be obtained, and subsequently the maximum profit which is the core of the land allocation in each AEZ can be calculated. Under the specific assumption of the functional form of the yield, the share of land use allocated to the land use type i is given with a logit share equation as follows:

$$S_{i,l,m} = \frac{\overline{\pi r}_{i,l,m}^{\frac{1}{\lambda}}}{\Sigma_p \overline{\pi r}_{i,l,m,p}^{\frac{1}{\lambda}}}, \tag{3}$$

Where λ is a positive parameter that represents the percentage change of land use in response to the change of the profit rate and $\overline{\pi r}_{i,l,m,p}^{\frac{1}{\lambda}}$ is the average profit rate of the land use type i, which is evaluated with the average or intrinsic yield $\overline{y_i}$ of the land use type i.

The land area of a specific land use type is calculated based on the total land area and the share of land use as follows:

$$\text{Landuse}_{i,l,m} = S_{i,l,m} \cdot \text{Totalland}_l,. \tag{4}$$

GTAP-AEZ Model

The GTAP-AEZ model is based on the GTAP-E model, an extended version of the GTAP model [26], which allows for the substitution between the capital and energy and that between various fuels in sector production (Figure 2). The equilibrium solutions were worked out using the GTAP-AEZ model, with the potential future economic activities assumed based on the RCP 4.5 scenario. The inputs into the production include capital, labor, land, and other intermediate inputs, and the GTAP-AEZ model takes full account of substitutions among the input factors such as the capital, labor, and land. Besides, there is a unique production function for each land use sector in AEZs in the GTAP-AEZ model, which reflects the difference of the productivity between different AEZs. For example, the productivity of the paddy rice production sector in AEZ i is different from that in AEZ j [27, 28], and this difference can be reflected with the production function of the paddy rice production sector. Nevertheless, it is assumed that all the paddy rice sectors in the same AEZ have a homogenous productivity and ability to meet market demand.

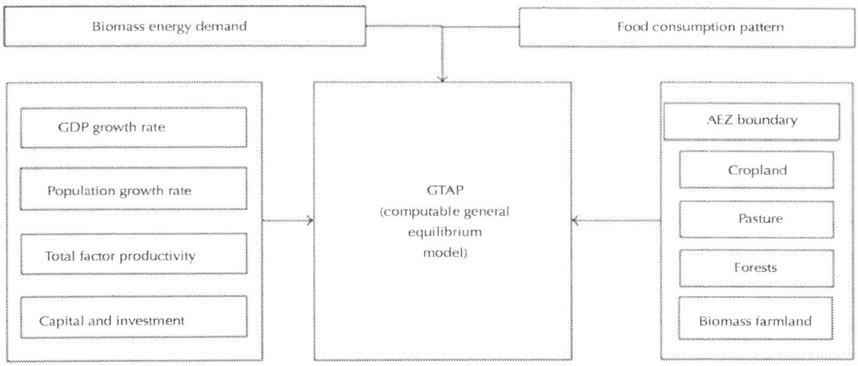

Figure 2: Framework of the GTAP-AEZ model.

We assume that transfer of land in a specific AEZ can occur between sectors, for which the land is more appropriate, and we believe that the introduction of the AEZ can render a sound presentation of sector competition for land. This is a new assumption beyond the standard GTAP model, in which it is assumed that the land is only transformable between sectors of crop production, livestock breeding, or timber plantation, regardless of climatic or soil constraints. However, facts show that most plants can only grow under the condition of certain temperature, moisture, soil types, land form, and so forth.

The GTAP-E model is a multisector, multiregion, and recursive dynamic computable general equilibrium model that extends the standard GTAP model through including the international capital mobility, endogenous capital accumulation, and the adaptive expectations of investment. This model is distinguished for its disequilibrium mechanism of determining the regional supply of investments. This disequilibrium mechanism includes the adjustment of the expected rate of return toward an actual return rate within each region and adjustment of the regional expected return rate toward the global return rate. These lagged adjustment mechanisms, as well as the mechanism of determining the composition of capital and allocation of wealth, are parameterized according to the econometric estimation documented by Golub (2006) [29]. In the analysis of the equilibrium of land use, it is assumed that the land is distributed among sectors for the maximization of profits in each period with similar capital and labor, although the land use does not change rapidly.

We split the total sector land rent into 18 AEZs according to the AEZ-specific production shares derived from the data provided by the Center for Sustainability and the Global Environment (SAGE) [15] as follows:

$$L_{ca} = L_c \cdot \left[\sum_{i \in SAGELANDUSE = c} p_i \cdot \frac{Q_{ia}}{H_{ia}} \cdot \frac{H_{ia}}{\sum_{a \in AEZS} \sum_{i \in SAGELANDUSE = c} P_i \cdot \frac{Q_{ia}}{H_{ia}} \cdot H_{ia}} \right]$$

$$c \in LANDUSE; i \in SAGELANDUSE; a \in AEZS, \qquad (5)$$

Where L_{ca} is the land rent accrued to the land use sector c in AEZ a; L_c is the land rent of the land use sector c, with no AEZ distinction P_i; is the per-ton price of SAGE's land use type i; Q_{ia} is the production (ton) of SAGE's land use type i in AEZ ; and H_a is the harvest area of SAGE's land use type i in AEZ. a The $\Sigma_{i \in SAGELANDUSE}$ operator means to aggregate over the disaggregated land use type i to the corresponding aggregated land use type. c Note that we assume that the per-ton land production price (P_i) is homogenous across the AEZs.

SCENARIOS

The Integrated Assessment Models (IAMs) explored a range of technological, socioeconomic, and policy futures that could lead to particular concentration pathways and magnitudes of climate change, which is represented by the RCPs. The RCPs include four different scenarios (Table 1), that is, one mitigation scenario leading to a very low forcing level (RCP2.6), two medium stabilization scenarios (RCP4.5/RCP6), and one very high baseline emission scenario (RCP8.5), all of which could be obtained with different combinations of economic, technological, demographic, policy, and institutional futures. The development of the RCPs in the first phase allows climate modelers to proceed with experiments in parallel to the development of emission and socio-economic scenarios, expediting the overall scenario development process [18]. Coupled carbon-cycle climate models can then as well calculate associated emission levels (which can be compared to the original emissions of the IAMs) [34].

Table 1: Description of RCPs

	Description	Publication-IA Model
CP8.5	Rising radiative forcing pathway leading to 8.5W/m2 in 2100	MESSAGE [30]
RCP6	Stabilization without overshoot pathway to 6W/m2 at stabilization	AIM [31]
RCP4.5	Stabilization without overshoot pathway 4.5W/m2 at stabilization after 2100	GCAM [32]
RCP2.6	Peak in radiative forcing at ~3W/m2 before 2100 and decline	IMAGE [33]

Two important characteristics of RCPs are reflected in their names. The word "representative" indicates that each of the RCPs represents a large set of scenarios in the various literature. In fact, as a set the RCPs should be compatible with the full range of emissions' scenarios available in the various current scientific literature, with and without the climate policy. The words "concentration pathway" means to emphasize that these RCPs are internally consistent sets of projections of the components of radiative forcing that are used in subsequent phases rather than the final new and fully integrated scenarios; that is, they are not a complete package of socio-economic, emission, and climate projections. The use of the word "concentration" instead of "emissions" also emphasizes that concentrations are used as the primary product of the RCPs and designed as inputs to climate models [35].

The RCP4.5 scenario is used in this study, under which the land use change is simulated with GCAM. Besides, the GTAP-AEZ model, which is similar to GCAM, is also used to analyze the land use structure in AEZs, and the results obtained with the two models were finally compared. The RCP4.5 scenario is a stabilization scenario in which the total radiative forcing is stabilized shortly after 2100, without overshooting the long-run radiative forcing target level [36]. RCP4.5 includes long-term, global emissions of greenhouse gases, short-lived species, and land use-land cover in a global economic framework which stabilizes the radiative forcing at 4.5 Watts per square meter (W/m^2), approximately 650 ppm CO_2-equivalent in the year 2100 without ever exceeding that value. The defining characteristics of this scenario are enumerated in Moss' papers [17, 18]. RCP 4.5 was updated from earlier GCAM scenarios to incorporate the historical emissions and land cover information and follows a cost-minimizing pathway to reach the target radiative forcing. The necessity to limit emissions in order to reach this target leads to the changes in the energy system, including shifts to electricity, lower emissions energy technologies, and the deployment of carbon capture and geologic storage technology. In addition, the RCP4.5 emission price is also applicable to the land use emissions. The simulated future emissions and land use were downscaled from the regional scale to the grid scale to facilitate the transfer to climate models. While there are many alternative pathways to achieve a radiative forcing level of 4.5 W/m^2, the application of the RCP4.5 provides a common platform for climate models to explore the response of the climate system to stabilize the anthropogenic components of radiative forcing.

RESULTS AND ANALYSIS

The results indicate that the land use areas in different AEZs, which are obtained with the GCAM model and the GTAP-AEZ model, are generally consistent (Figure 3). The pastureland areas simulated with the two models differ most greatly, but they are still generally consistent in different AEZs. Besides, the results obtained with the GTAP-AEZ model and the GCAM model both show that the grassland is approximately equally distributed in different AEZs, but the grassland area in different AEZs differs a bit more greatly in the result obtained with the GCAM model. In addition, the results obtained with the two models show that the forestland is mainly located in AEZ9-AEZ12, while the shrubland and cropland are mainly in AEZ7-AEZ13. What is more, the built-up land, the area of which is the smallest, is generally distributed in AEZ10. Through comparing the results obtained with the two models, it is found that the distribution of different land use types among AEZs is approximately consistent, but with some difference in the total areas of different land use types.

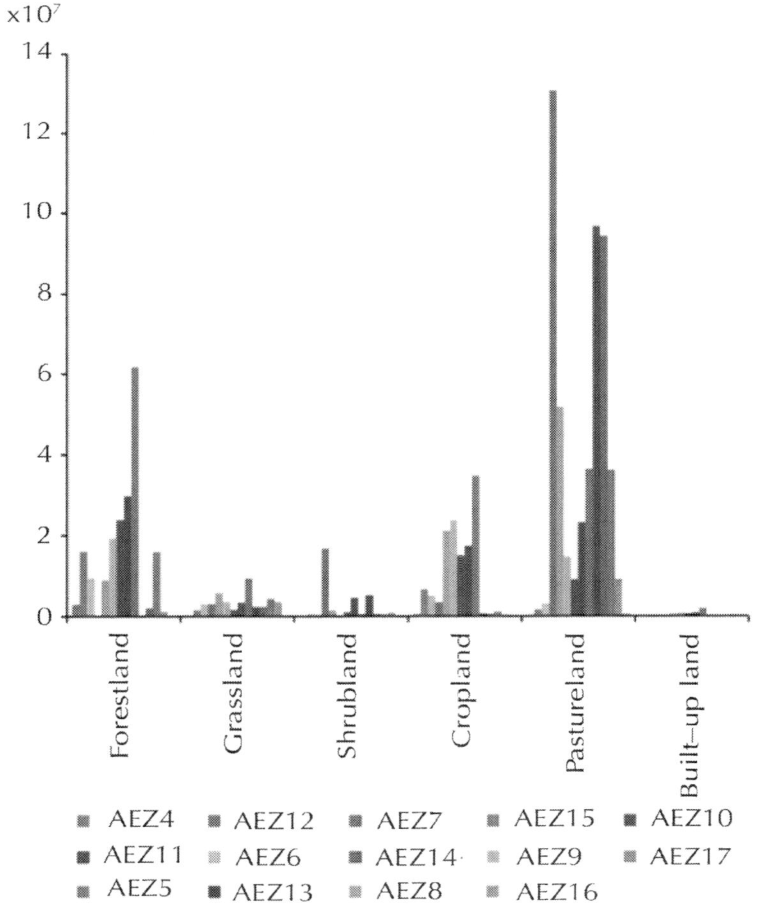

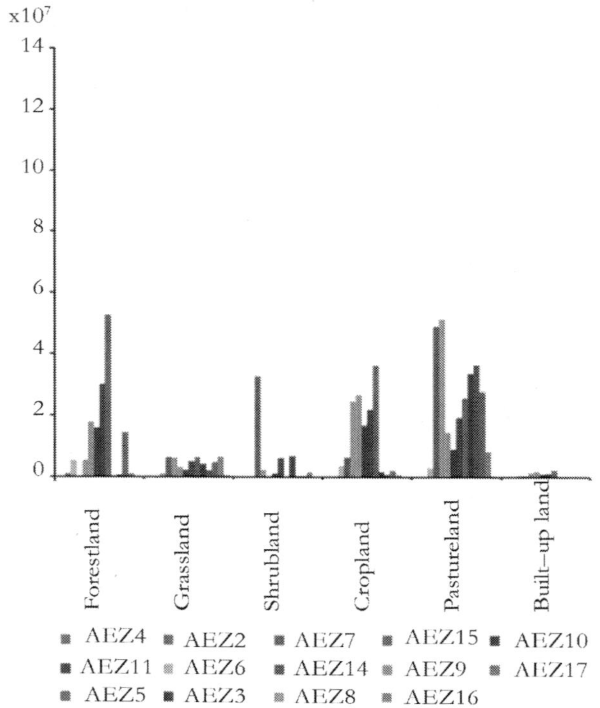

Figure 3: Simulated land use area in 14 AEZs in 2010 using the GCAM model (a) and GTAP-AEZ model (b) (unit: hectare).

There is still some difference between the real land use area and that obtained with the two models, and the simulation result with the GTAP-AEZ model is better than that with the GCAM model (Figure 4). The result shows that the areas of cropland and forestry land simulated with the GCAM model and the GCAM model are far higher than the real one. The area of cropland obtained with the GCAM model and the GTAP-AEZ model is 2.13 and 1.96 times larger than the real one, respectively. However, the areas of grassland and built-up land simulated with the two models are both lower than real values. This indicates that there is still some inaccuracy in data of the land use structure, industry structure, and social economic situation of China in the global simulation. For example, the forest land should be divided into economic forest lands and ecological forest lands, but the forest land was not distinguished in this study, which leads to the significant difference between the simulated and real areas of the forest land.

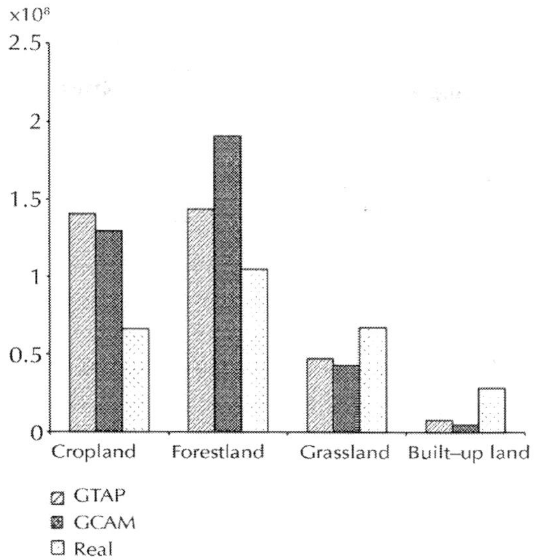

Figure 4: Comparison between the simulated and real land use areas in 2010 (unit: hectare).

There are an extremely complex socioeconomic structure and land use structure in China, both of which have changed greatly due to the rapid economic development of China and consequently made it very difficult to accurately simulate the land use change in China using models with static data. For instance, the estate price in China has fluctuated prominently during the past decades, but a static estate price was first used in the GCAM model in this study, which makes it obviously difficult to simulate the constantly changing industrial structure in China. Therefore, it is necessary to calibrate the models before simulating the land use change.

In order to more accurately simulate the change of the land use structure in China according to the reality of China and improve the precision of future scenario simulation, we calibrated the parameters of the GCAM model and the GTAP-AEZ model (Table 2). The influence of policy intervention was included in the models according to the specific national condition of China, and other parameters were also calibrated. In this study, the price of agricultural products was set to increase by 1.5 percent every year; TFP will increase by 0.1 percent, and the annual population growth rate will decrease from 0.8% to

0.5%. The results indicate that the land use structure simulated with the calibrated GCAM model becomes much more accurate than it was and has more closely approached to that in the reality. Besides, the simulation accuracy with the calibrated GCAM model is much higher than that with the calibrated GTAP-AEZ model (Figure 5).

Table 2: The adjustment of GCAM input parameters in 2010 (unit: %)

Input parameters	Previous parameters	Adjusted parameters
GDP growth	10	10.4
Labor growth	0.4	0.4
Capital growth	12.6	12.8
TFP growth	0.9	1.1
Population growth rate	0.8	0.5

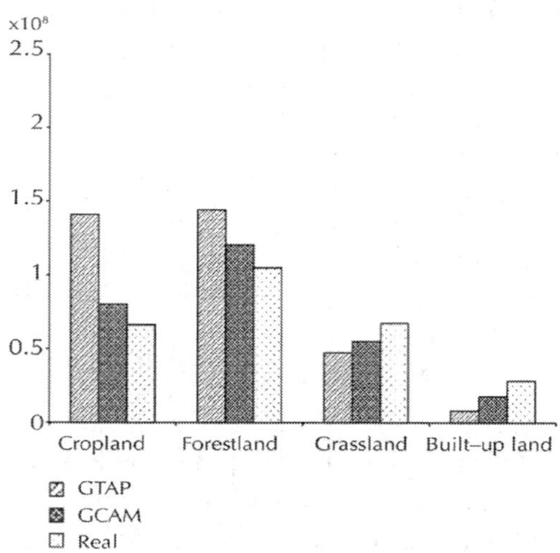

Figure 5: The comparison of land use area in China in 2010 simulated with the calibrated GCAM model, the GTAP model, and the real values (unit: hectare).

CONCLUSIONS AND DISCUSSION

This study simulated the land use change in China under the RCP 4.5 scenario with GCAM and GTAP-AEZ models and compared the simulated and real land use structures. Overall, the simulation results obtained with GCAM and GTAP-AEZ are consistent, but also with some difference, and the land use structure simulated with GTAP-AEZ is to be more close to the real conditions in different AEZs than that obtained with GCAM. For example, the consistence between the forest land area simulated with GCAM and the real one reached more than 80%, while that with GTAP-AEZ reached only 37%. GCAM involves the driving factors of the rapid economic development, which makes the simulation more close to the reality. However, neither of the two models takes account of the impacts of policies on socioeconomic development, which also have great influence on the land use change. Therefore, it is necessary to calibrate the models through optimizing the model input parameters. When the models are calibrated through adjusting these socioeconomic parameters according to the specific conditions, the overall simulation accuracy of GCAM reached 82%, while that of GTAP-AEZ also reached 60%, indicating that it is possible and necessary to improve the simulation accuracy through optimizing input parameters of the models according to the specific conditions.

In recent decades, more and more land use simulation models have developed, but it is still a hard task to implement the calibration of input parameters of these model. In the study, the land use structure of China in 2010 under the RCP 4.5 climate change scenario was simulated with GCAM and GTAP-AEZ, both of which were further calibrated through adjusting the input parameters, focusing on comparing simulation accuracy of the two models. The result indicates that the areas of cropland and forest land simulated with the two models were higher than the real one, while the simulated areas of grassland and built-up land were lower than the real values, and the simulation accuracy was greatly improved after the model calibration. However, due to the uncertainties of climate change, economic development, and other factors, it is very difficult to accurately simulate the long-term land use change in the future, and therefore it is necessary to implement more in-depth research on how to optimize the parameters according to the specific conditions in the future.

ACKNOWLEDGMENTS

This research was supported by the Key Project funded by the National Key Programme for Developing Basic Science in China (Grant no. 2010CB950900), National Natural Science Foundation of China (Grant no. 41071343), and the Key Projects in the National Science & Technology Pillar Program (Grant no. 201311001-5).

REFERENCES

1. P. M. Vitousek, H. A. Mooney, J. Lubchenco, and J. M. Melillo, "Human domination of Earth's ecosystems," Science, vol. 277, no. 5325, pp. 494–499, 1997.
2. X. Deng, C. Zhao, and H. Yan, "Systematic modeling of impacts of land use and land cover changes on regional climate: a review," Advances in Meteorology, vol. 2013, Article ID 317678, 11 pages, 2013.
3. E. F. Lambin, X. Baulies, N. E. Bockstael, et al., "Land-use and land-cover change (LUCC), implementation strategy," IGBP Report 48, IHDP Report 10, IGBP, Stockholm, Sweden; IHDP, Bonn, Germany, 2000.
4. B. L. Turner II, "Local faces, global flows: the role of land use and land cover in global environmental change," Land Degradation & Rehabilitation, vol. 5, no. 2, pp. 71–78, 1994.
5. J. Y. Liu and X. Z. Deng, "Progress of the research methodologies on the temporal and spatial process of LUCC," Chinese Science Bulletin, vol. 55, no. 14, pp. 1354–1362, 2010.
6. S. Cao, C. Wu, and W. Yu, "Evaluation of land ecological service and its application in overall arrangement of land use," Bulletin of Soil and Water Conservation, vol. 20, no. 2, pp. 197–200, 2006.
7. Q. Jiang, X. Deng, H. Yan, D. Liu, and R. Qu, "Identification of food security in the mountainous guyuan prefecture of China by exploring changes of food production,"Journal of Food, Agriculture & Environment, vol. 10, no. 1, pp. 210–216, 2012.
8. X. Deng, F. Yin, Y. Lin, Q. Jin, and R. Qu, "Equilibrium analyses on structural changes of land uses in Jiangxi Province," Journal

of Food, Agriculture & Environment, vol. 10, no. 1, pp. 846–852, 2012.

9. Y. Hu, Y. Liu, and X. Deng, "Relativity analysis on land use and land cover change and optimal allocation of land resources," Progress in Geography, vol. 23, no. 2, pp. 51–57, 2004.
10. A. Flamenco-Sandoval, M. Martínez Ramos, and O. R. Masera, "Assessing implications of land-use and land-cover change dynamics for conservation of a highly diverse tropical rain forest," Biological Conservation, vol. 138, no. 1-2, pp. 131–145, 2007.
11. X. Deng, H. Su, and J. Zhan, "Integration of multiple data sources to simulate the dynamics of land systems," Sensors, vol. 8, no. 2, pp. 620–634, 2008.
12. Y. Cai, Y. Liu, Z. R. Yu, and P. H. Verburg, "Progress in spatial simulation of land use change-CLUE-s model and its application," Progress in Geography, vol. 23, no. 4, pp. 63–71, 2004.
13. J. Y. Liu and X. Z. Deng, "Progress of the research methodologies on the temporal and spatial process of LUCC," Chinese Science Bulletin, vol. 55, no. 14, pp. 1354–1362, 2010.
14. A. Brenkert, S. Kim, A. Smith, and H. Pitcher, "Model Documentation for the MiniCAM," U.S. Department of Energy, PNNL-14337, 2003.
15. H. Lee, Incorporating Agro-Ecologically Zoned Land Use Data and Land-Based Greenhouse Gases Emissions into the GTAP Framework, Centre for Global Trade Analysis, Purdue University, West Lafayette, Ind, USA, 2005.
16. J. M. Burniaux and H. L. Lee, "Modelling land use change in GTAP," 2005.
17. R. Moss, M. Babiker, S. Brinkman, and E. Calvo, "Towards new scenarios for analysis ofemissions, climate change, impacts, and response strategies," IPCC Expert Meeting Report, Intergovernmental Panel on Climate Change, Geneva, Switzerland, 2008.
18. R. H. Moss, J. A. Edmonds, K. A. Hibbard et al., "The next generation of scenarios for climate change research and assessment," Nature, vol. 463, no. 7282, pp. 747–756, 2010.
19. M. Wise, K. Calvin, A. Thomson et al., "Implications of limiting CO_2 concentrations for land use and energy," Science, vol. 324, no. 5931, pp. 1183–1186, 2009.

20. J. Edmonds, M. Wise, H. Pitcher, R. Richels, T. Wigley, and C. MacCracken, "An integrated assessment of climate change and the accelerated introduction of advanced energy technologies: an application of MiniCAM 1.0," Mitigation and Adaptation Strategies for Global Change, vol. 1, no. 4, pp. 311–339, 1997.
21. J. Edmonds and J. Reilly, Global Energy: Assessing the Future, Oxford University Press, New York, NY, USA, 1985.
22. R. D. Sands and J. A. Edmonds, "Climate change impacts for the conterminous USA: an integrated assessment—part 7: economic analysis of field crops and land use with climate change," Climatic Change, vol. 69, no. 1, pp. 127–150, 2005.
23. R. D. Sands and M. Leimbach, "Modeling agriculture and land use in an integrated assessment framework," Climatic Change, vol. 56, no. 1-2, pp. 185–210, 2003.
24. T. M. L. Wigley and S. C. B. Raper, "Implications for climate and sea level of revised IPCC emissions scenarios," Nature, vol. 357, no. 6376, pp. 293–300, 1992.
25. M. Hulme, T. Jiang, and T. M. L. Wigley, SCENGEN: A Climate Change Scenario Generator, a Software User Manual, Climatic Change Research Unit, School of Environmental Sciences, University of East Anglia, Norwich, UK, 1995.
26. J. M. Burniaux and T. Truong, "GTAP-E: an energy-environmental version of the GTAP model," GTAP Technical Paper 16, Center for Global Trade Analysis, Purdue University, West Lafayette, Ind, USA, 2001.
27. E. Ianchovichina and R. McDougall, "Structure of dynamic GTAP," GTAP Technical Paper 17, Center for Global Trade Analysis, 2001, http://www.gtap.org/.
28. X. Deng, J. Han, and F. Yin, "Net energy, CO_2 emission and land-based cost-benefit analyses of Jatropha biodiesel: a case study of the Panzhihua region of Sichuan province in China," Energies, vol. 5, pp. 2150–2164, 2012.
29. A. Golub, Projecting the global economy to 2025: a dynamic general equilibrium approach [Ph.D. thesis], Purdue University, 2006.
30. K. Riahi, A. Grübler, and N. Nakicenovic, "Scenarios of long-term socio-economic and environmental development under climate

stabilization," Technological Forecasting and Social Change, vol. 74, no. 7, pp. 887–935, 2007.

31. Y. Hijioka, Y. Matsuoka, H. Nishimoto, T. Masui, and M. Kainuma, "Global GHG emission scenarios under GHG concentration stabilization targets," Journal of Global Environment Engineering, vol. 13, pp. 97–108, 2008.
32. S. J. Smith and T. M. L. Wigley, "Multi-gas forcing stabilization with minicam," The Energy Journal, vol. 27, pp. 373–391, 2006.
33. D. P. van Vuuren, B. Eickhout, P. L. Lucas, and M. G. J. den Elzen, "Long-term multi-gas scenarios to stabilise radiative forcing—exploring costs and benefits within an integrated assessment framework," Energy Journal, vol. 27, pp. 201–233, 2006.
34. K. A. Hibbard, G. A. Meehl, P. M. Cox, and P. Friedlingstein, "A strategy for climate change stabilization experiments," Eos, vol. 88, no. 20, pp. 217–221, 2007.
35. F. Wu, X. Deng, F. Yin, and Y. Yuan, "Projected changes of grassland productivity along the representative concentration pathways during 2010–2050 in China," Advances in Meteorology, vol. 2013, Article ID 812723, 9 pages, 2013.
36. J. Liu, J. Zhan, and X. Deng, "Spatio-temporal patterns and driving forces of urban land expansion in China during the economic reform era," Ambio, vol. 34, no. 6, pp. 450–455, 2005.

Chapter 4

Modeling Agro Ecosystem Services under Simulated Climate and Land-use Changes

Abdullah A. Jaradat[1] and George Boody[2]

[1]USDA-ARS, 803 Iowa Avenue, Morris, MN 56267, USA

[2]Land Stewardship Project, 821 E 35th Street, Minneapolis, MN 55407, USA

ABSTRACT

Ecological functioning of the intensive, homogeneous agro ecosystems in the Chippewa River Watershed (CRW), MN, USA, can be improved by reducing soil erosion, runoff, and nutrient leaching. These ecosystem services can be achieved through increased perennials in crop rotations to diversify land use and sustain carbon sequestration. We calibrated, validated, and used APSIM software to simulate the effect of 100 yrs. each of historical and future climate change scenario (IPCC-A2) on biophysical processes in representative soil types of the predominant farming systems in CRW. The interrelationships between crop rotations, soil types, climate variables, and ecosystem services indicated that not all objectives of sustainable agro-ecosystem are compatible, and tradeoffs among them are necessary. Site-specific and diversified crop rotations that comply with the environmental constraints of climate and soils could lead to more efficient implementation of strategies to improve ecosystem services in the watershed if current management practices of high external inputs and tillage persisted.

INTRODUCTION

The extent and global scale of the environmental costs associated with intensive agriculture are already documented [1], and their effects on ecosystem services have been extensively researched [2, 3]. Ecological functioning of the intensive agro ecosystems in the Upper Midwest, in general [4], and in MN's Chippewa River Watershed (CRW), in particular, is being compromised by soil erosion, runoff, and nutrient leaching [5, 6]. The land use in CRW is predominately centered on commodity production systems that deliver corn, soybean, and livestock products for domestic consumption and for an expanding export market [6, 7]. Similar to other conventional land-use systems in the Upper Midwest of the Corn Belt (i.e., corn-soybean rotation, conventional tillage, N fertilizer), losses of ~30% of the average 125 kg applied nitrogen (N) per ha to subsurface drainage in CRW are not uncommon [7, 8]. The rise in intensive, homogeneous croplands, especially corn production systems, is responsible for a substantial proportion of the increase in NO_3-N values observed in surface and underground waters of the Midwest since the 1970s. River-born nutrients, especially NO_3-N from tile-drained fields in the CRW, contribute to environmental pollution within and beyond the watershed [9]. Increased corn production as a feedstock for ethanol would result in a concomitant increase in N loads of ~5 to 18% [10]. This raises concerns about the environmental, agronomic, and economic sustainability of the corn-soybean crop rotation in this and similar watersheds in the Upper Midwest.

Recent research [11] indicated that managed ecosystems often fail to respond smoothly to external changes. Therefore, more efforts are being devoted to studying ecological regime shifts, thresholds, and resilience of ecosystems. This is becoming ever more important for agro ecosystems because many rural communities depend on their provisioning services. These services are primarily determined by the structural diversity (i.e., number of crops within a field) and temporal diversity (i.e., complex crop rotations), and their interaction with management and climatic factors, and contribute to diversification of agro ecosystems. However, the antagonistic relationship between ecosystems' capacity to simultaneously sustain the availability of regulating services and agricultural production of an agro-ecosystem should be considered when designing new management practices and complex crop rotations [12].

The CRW supports agro ecosystems that flourish in a temperate climate regime that could see significant changes in the next few decades due to climate [13] and land-use changes [13, 14]. This watershed is a source of many ecosystem services including provisioning services such as the production of food, feed, and biofuel; regulating services such as climate regulation, flood regulation, and water quality control; cultural services such as recreational, aesthetic, and other benefits; supporting services such as nutrient cycling [3]. This watershed is among the most altered landscapes in the Upper Midwest. The majority (60–94%) of the land area within the watershed is used for row crops and livestock production, and pressure to increase row crop production is intensifying with elevated demand for biofuels [10]. Fertilizer application, erosion attributed to mechanical tillage and exposed soil, and rapid movement of nutrients and sediment to streams via tile drainage are primary causes for high nitrate and turbidity in this [15] and other watersheds in the Midwest [14].

Historical land-use changes from prairie native grasses to row crops have had a significant effect on the hydrology of the CRW [4, 5]. It is anticipated that future population growth and changing agricultural demands will inevitably force land-use changes and, therefore, land cover in the watershed [15]. Limited land cover during the short growing season (~110 days yr^{-1}) by annual crops in the CRW plays a significant role in influencing the water and energy balance at the land surface via its effect on light interception, transpiration, and evaporation from crop canopies. In addition to the physical properties of soil types in the watershed that can have a strong influence on the nature of the hydrologic response, changes in vegetation cover can alter the surface energy balance and evapotranspiration. Land-use change from native prairie in the CRW to row crops resulted in about 10–15% decrease in evapotranspiration and increases of 20–30% in total runoff [16].

Results of earlier simulation studies [15] suggested that average annual percent change in precipitation, runoff, soil loss, and crop yield, in response to projected climate change in six locations in MN, would increase by 16.1, 34.6, 22.5, and 58.6%, respectively. Similar projections in biophysical processes, except crop yield, were obtained [17] using historical and modeled data in the Midwest. Earlier monitoring efforts in the CRW [5] resulted in establishing a water quality and quantity baseline and emphasized the need for more perennials on the landscape in order to enhance both environmental and agronomic agro-ecosystem services, promote conservation incentives, and develop viable markets for perennials and integrated crop-livestock production systems. Previous modeling efforts predicted that land resources in the CRW can be multifunctional; they can provide a large number of functions related to social, economic, and environmental prosperity and sustainability [4, 6]. Given the complexity of agro ecosystems, and the spatiotemporal variability of the involved biophysical processes within the CRW, modeling is needed for assessing the effect of climate change and alternative management strategies [18]. These strategies should include more permanent plant cover on the landscape and their effect should be assessed on a number of ecosystem services cum biophysical processes (e.g., biomass and grain yield of crops, carbon sequestration in the soil, nitrogen loss, runoff, and soil erosion).

Due to current land-use systems, the CRW faces a number of environmental challenges, including degradation of water quality, threats to biodiversity, and increased flooding, soil erosion, and nutrient loss to both surface and underground water [4, 19]. In addition to these environmental challenges, the effect of spatiotemporal climate variability in the watershed is likely to be intensified by climate change, which is predicted to disrupt many ecosystem functions, altering their capacity to provide goods and services and rendering them more susceptible to degradation [20, 21]. The question facing farmers and researchers alike is how to optimize a diversified agro-ecosystem, maximize production and profits, and minimize negative environmental effect of future agroecosystems [11]. The objectives of this simulated study were to (1) predict the long-term effect of diverse crop rotations on agro-ecosystem services and the relationship between soil types, crop rotations, and their response to historical and future climate change, (2) model the effect of targeted land-use changes on managed agroecosystems, and (3) develop predictive and validation models of future ecosystem services under the A2 climate change scenario if current management practices of high external inputs and conventional tillage are continued.

MATERIALS AND METHODS

Watershed Physiography

The Chippewa River drains 5,387 km^2 of mixed natural and managed ecosystems in West Central MN (Figure 1). Annual crops, predominantly grown in monocultures, occupy between 60 to 94% of the land area in the major agro-ecological regions of the watershed [19]. Recently, however, increased demand for domestic biofuels is expected to influence land-use options [22] and will impact several biophysical processes in managed and natural agro ecosystems within the watershed. The CRW has high-value ecological features including seven major lakes, two state parks with prairie, forest and lake areas, numerous wildlife management and waterfowl production areas, and ~3,000 km of intermittent and perennial streams [4–6].

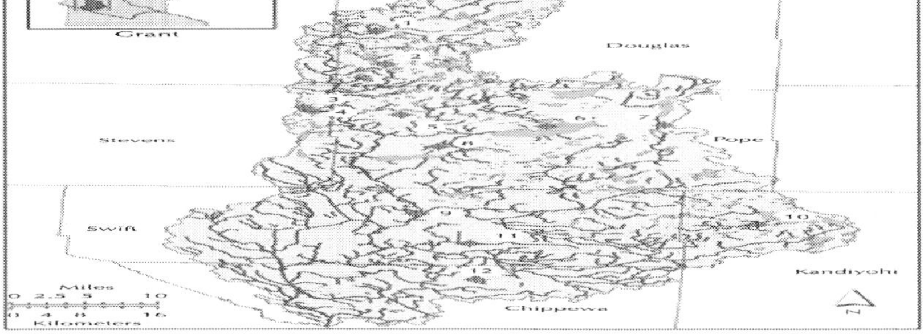

Figure 1: Location of the Chippewa River Watershed in south-central MN and the 12 soil type sites selected for a simulation study under past and future climate change scenarios.

Selection of Soil Types

Given the complexity of the agro ecosystems and the spatiotemporal variability of the involved biophysical processes in the CRW, we selected 12 representative soil types (series) to cover the major physiographic features of the watershed (Figure 1). General information on these soils is available athttp://soils.usda.gov/survey/geography/ssurgo/ detailed characteristics can be found athttp://soils.usda.gov/survey/geography/ssurgo/description.html. The soil types (series) were (1) BaB2: (408 meters above sea level, m) Evansville Barnes loam, 2–6% slope, eroded; (2) Fa: (411 m) Kensington Flom silty clay loam, 0–2% slope; (3) BaB: (369 m) Swan Lake Barnes loam, 3% slope; (4) HaA: (361 m) Swan Lake Hamerly clay loam, 0–3% slope; (5) SuA: (383 m) Cyrus Svea loam, 0–2% slope; (6) BbC2: (356 m) Long Beach Barnes-Langhei loams, 6–12% slope, eroded; (7) EvB: (415 m) Glenwood Estherville loam, 2–6% slope; (8) Pf: (365 m) Starbuck Parnell and Flom silty clay loams; (9) J55A: (314 m) Benson Sedgeville loam, channeled, 0–2% slope, occasionally flooded; (10) L33B: (373 m) Sunburg Kandiyohi clay, 2–5% slope; (11) J51A: (312 m) De Graff Bearden-Quam, depression, complex, 0–2% slope; (12) J30A: (347 m) Clara City Tara silt loam, 1–3% slope.

Ecosystem Services

Data on a number of ecosystem services (or biophysical processes, i.e., total biomass and grain yield of crop rotations, carbon sequestration in the top 1 m soil profile, NO_3-N and NH_4-N loss, runoff, and soil erosion; Table 1) were compiled from simulated data to quantify and evaluate the long-term effects of climate change, soil types, traditional corn-soybean crop rotation, and more complex crop rotations with corn-soybean-wheat and a forage perennial crop, and their interactions in the watershed.

Table 1: Results of univariate analysis of variance of ecosystem services based on simulation data in response to climate, soil series, and their interaction in the Chippewa River Watershed in MN, USA

sEcosystem service	Statistic	Climate (past, future)	Soil series	Soil series × climate
	d.f.	1	11	11
Biomass	F-value	122	99	13
	Probability	<0.001	<0.001	<0.01
Grain yield	F-value	232	88	14
	Probability	<0.001	<0.001	<0.01
C-sequestration	F-value	1242	15322	1654
	Probability	<0.001	<0.001	<0.001
NO3-N	F-value	8.5	72	14
	Probability	<0.05	<0.01	<0.05
NH4-N	F-value	1591	719	24
	Probability	<0.001	<0.001	<0.001
Runoff	F-value	23038	4259	1472
	Probability	<0.001	<0.001	<0.001
Erosion	F-value	257	2037	621
	Probability	<0.001	<0.001	<0.001

Simulation Scenarios and Modeling

Each of one hundred and sixty eight combinations of (1) two global climate change scenarios (Past (P-), based on historical weather local data for the period 1900–1999, and future (F-), based on simulated weather data for the period 2000–2099, climate change scenarios; see below), (2) seven crop rotations (corn-soybean, corn-soybean-wheat-(1–5 years of alfalfa), and continuous alfalfa for eight years), and (3) twelve soil types was simulated for 100 years. We utilized the Agricultural Production Systems Simulator (APSIM 7.3; 23) that integrates modules of cropping, management, and biophysical processes within farming systems for improved decision support. APSIM is a point-based model that runs on a daily time step and is capable of yield prediction across land areas. It estimates crop productivity and potential effects of farming systems at selected locations in the study area based on the site's soil properties and climate regimes [23]. The soil nitrogen (NO_3-N and NH_4-N) and carbon transformation within APSIM are largely based on experiences within the CERES model; both variables were estimated to a depth of 1 m in the soil profile. APSIM includes a generic crop module to capture physiological characteristics and processes that are shared between different crop species. The curve number method [24], where soil type, land use, management practices, and antecedent moisture are lumped and represented on average by a curve number, was used to simulate runoff. The conceptual basis of this method is supported by empirical data. Soil erosion was estimated using runoff volume, vegetative cover, land slope, and several soil physical and chemical properties [23].

A database derived from on-station and on-farm research on traditional and alternative cropping systems, including perennial biomass crops, was utilized to calibrate the simulation model. Sensitivity analysis was carried out for each crop in the crop rotation. The database included detailed quantitative measurements on crops, soils, nutrients, and weather variables [25]. To conduct the simulations and subsequent multivariate statistical analyses, we used historical weather data for the past 100 years (compiled from the weather station of West Central Research and Outreach Center, Morris, MN, and will be made available upon request at \\Mw45-ws1\Common Data Weather Data\WCROC 8AM observations (1886 - present)\WCROC Weather1886-2010.xlsx) and will be referred to as the P-climate change scenario, as well as simulated changes in future (21st century) temperature, precipitation, and CO_2 concentrations based on the A2 climate change scenario which was generated by the Coupled Global Climate Model (CGCM) 3.1 which was developed by the Canadian Center for Climate Modeling and Analysis, Canada (http://www.ec.gc.ca/ccmac-cccma/default.asp?lang=En) and will be referred to as the F-climate change scenario. The IPCC A2 scenario was selected for simulations (Figure 2) because it assumes "business as usual" in terms of CO_2 emissions and is characterized by a high rate of growth in CO_2 emissions and most closely reproduced the actual emissions trajectories during the period since the scenario was defined [20].

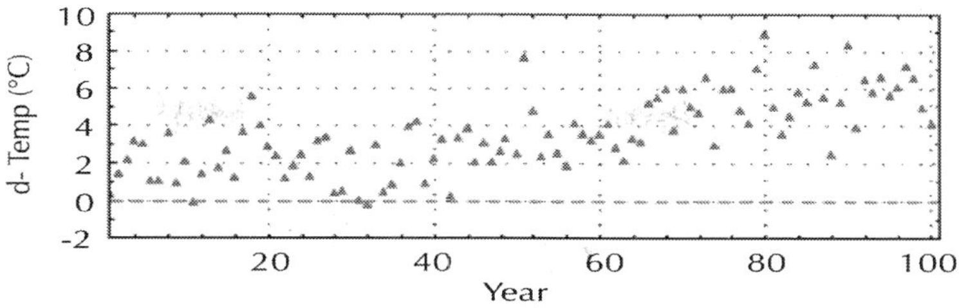

(a)

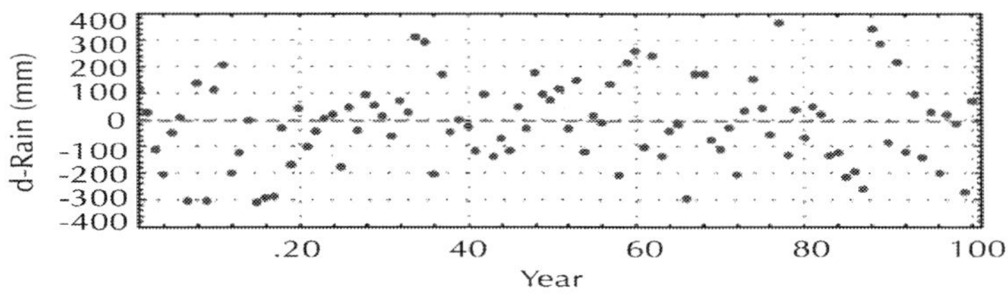

(b)

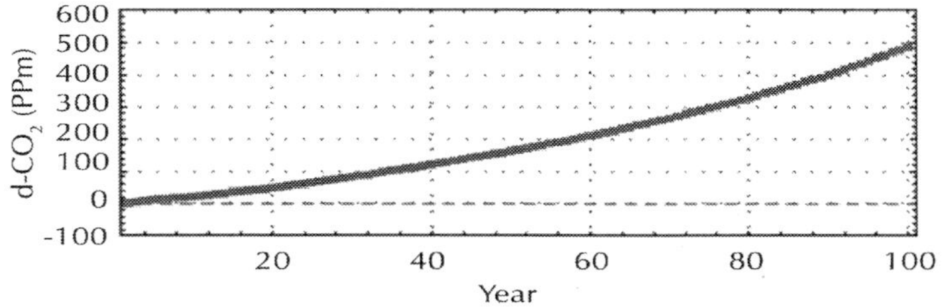

(c)

Figure 2: Deviations of projected mean temperature (d-Temp, °C), total rainfall (d-Rain, mm), and CO_2 concentration (d-CO_2, ppm) for the next 100 years based on the future (A2 scenario) climate change scenario from the respective mean values derived from historical records for the past 100 years (P1 scenario) in the Chippewa River Watershed, MN.

For each of the 12 locations, representing soil types, conventional management practices, as defined in Jaradat and Weyers [25], and position in the landscape within the watershed (Figure 1), we assessed the effect of current conventional (corn-soybean) and alternative crop rotations (corn-soybean-wheat-alfalfa (with 1 to 5 years of alfalfa following three years of grain crops), and continuous alfalfa for 8 years) on several crop and soil biophysical processes [26]. Conventional tillage and local management practices were used in simulations along with 100-year historical weather data and 100 years of projected weather data based on the A2 climate change scenario as predicted by CGCM 3.1 [20].

Statistical Analyses

Data derived from each simulation run were subjected to several statistical analyses procedures in order to (1) assess the effect of climate, soil types and their interaction on each ecosystem service using univariate analysis of variance procedures; (2) quantify the effect of, and estimate the variance due, to fixed (crop rotations and crops within crop rotations) and random factors (soils types and their interaction with crop rotation) on these ecosystem services taking the spatial and physiographic (longitude, latitude, and elevation) characteristics of the 12 soil types into consideration; and using the spatial option in the multivariate residual maximum likelihood (REML) procedure; (3) classify the soil types according to their reaction to climate change and identify which ecosystem services significantly contributed to this classification; (4) contrast ecosystem services based on past and future climates and develop predictive models for each ecosystem service; (5) quantify the effect of increased proportion of a perennial crop on biomass and grain yield of crop rotations and develop statistical models to predict its effect on the remaining ecosystem services; (6) develop partial least square regression models to assess the predictability of future status of ecosystem services based on their current status in each soil types. Annual data output was used to carry out statistical procedures described in (1) to (3), whereas normalized aggregate data were used to carry out the statistical procedures in (4) to (6). The largest value of each ecosystem service in all 12 soil types was considered as 1.0 and the remaining 11 values were expressed as a percent of that value. Normalized data is more appropriate to use for comparisons across all ecosystem services due to different units of measurement and to assess multiple agro-ecosystem functions [27]. Several modules in STATISTICA [28] and GenStat software programs [29] were used to carry out statistical analyses and to develop predictive models.

RESULTS

The 12 sites selected for the simulation study represented the predominant soil types, cropping systems, and physiographic features in the watershed (Figure 1). In addition, these locations represent different positions in the landscape with slopes ranging from 0.0 to 12%. The North-South and West-East spread of the 12 locations and their proximities/distances from water bodies as well as their soil physical and chemical properties and soil profile characteristics captured the maximum diversity within the watershed and provided necessary data to validate the simulation results.

Past and Future Climates

Total annual rainfall under P- and F-climate change scenarios were almost equal (611 and 596 mm, resp.); however, differences in minimum and maximum rainfall were large in magnitude and resulted in the respective ranges of 628 and 332 mm, coefficients of variation of 21.1 and 12.4%, skewness of 0.026 and −0.098, and kurtosis of −0.22 and −0.49. Mean annual temperatures and their standard deviations (°C) for P1 and A2 were 5.74 (1.07) and 9.35 (1.58). Although they had almost equal skewness values (0.205 and 0.25, resp.), their respective kurtosis values (0.12 and −0.36) were very different. Carbon dioxide (CO_2) is projected to increase (almost linearly from the initial level of 370 to 856 ppm at the end of the 21st century; Figure 2) under the A2 climate change scenario; however, it is expected to display a large level of variation (coefficient of variation = 25.2%) and will be positively skewed (0.47) with a negative kurtosis (−0.95) value.

Deviations of projected mean temperature (d-Temp, °C), total rainfall (d-Rain, mm), and CO_2concentration (d-CO_2, ppm) for the next 100 years based on the A2 climate change scenario from the respective mean values derived from historical records for the last 100 years (P-scenario) are presented in Figure 2. Future mean temperature is projected to be higher by 2–8°C, especially towards the second half of the 21st century. Deviations of future from past rainfall fluctuated widely (from −300 to +395 mm), whereas those of CO_2 are expected to range from 0.0 (at the start of the simulation period) to ~500 ppm (at the end of the 21st century) above the current level assuming "business as usual" in terms of CO_2 emissions [20]. d-Temp was negatively correlated ($r = -0.64$; $P < 0.001$) with past temperature (P1) and positively correlated with projected temperature ($r = 0.85$; $P < 0.001$) and CO_2 ($r = 0.71$; $P < 0.001$) under A2 climate change scenario. d- Rain exhibited a parallel trend; it was negatively correlated with past rainfall under P-scenario ($r = -0.88$; $P < 0.001$) and positively correlated with rainfall under A2 ($r = 0.57$; $P < 0.001$). d- CO_2, on the other hand, was positively correlated with future temperature under A2 ($r = 0.83$; $P < 0.001$) and with d-Temp ($r = 0.71$; $P < 0.001$).

Effect of Climate, Soil Types and Their Interaction on Ecosystem Services

Results of univariate analyses of variance of ecosystem services based on simulated response to climate change, soil types, and their interaction (Table 1) indicated that all ecosystem services are likely to be impacted by past and future climates, soil types, and their interaction. The F-values (Table 1) and their probabilities suggested that climate change would be the major factor in determining the future magnitude of biomass and grain yield production, the loss of NH_4-N, and runoff, whereas soil types would be the main determinant of the level of carbon sequestration, the loss of NO_4-N, and soil erosion. The magnitude of the effect of the interaction between climate and soil types, when compared with the effect of climate and soil types, would be a major contributor to determining future grain yield, runoff, and soil erosion levels.

Multivariate Analyses of Ecosystem Services

The variance portion attributed to fixed (crop rotations and crops within crop rotations) and random (soils types and their interaction with crop rotation) factors on ecosystem services taking into consideration the spatial and physiographic (longitude, latitude, and elevation) characteristics of the 12 soil types (Table 2) suggested that different factors and their interactions accounted for a wide range of variability in past and future ecosystem services. Crop rotations, whether based on past or future climates, differed significantly for all ecosystem services under study; however, crops within crop rotations differed significantly as to their biomass and grain yield under both climate change scenarios. The soil-related processes (soil carbon, NO_3-N, NH_4-N, runoff, and erosion) were not significantly impacted by differences between crops within crop rotations. Nevertheless, under future (A2) climate change scenario, all variables except soil carbon differed significantly from those based on past weather.

Table 2: Variance components and their level of significance accounted for by fixed factors, random factors, and their interaction in two simulation scenarios involving 12 soil series and eight crop rotations in the Chippewa River Watershed, MN

		Fixed factor		Random factor	
Simulation	Variable	Crop rotation	Crops (crop rotation)	Soil series	Crop rotation × soil series
		Probability of F-value		Probability of z-value (% variance)	

Past 100 yrs	Biomass	0.001	0.001	0.05 (61%)	0.21 (3%)
	Grain yield	0.005	0.05	0.02 (45%)	0.09 (6%)
	Soil carbon	0.0001	0.12	0.01 (74%)	0.05 (12%)
	NO3-N	0.005	0.15	0.05 (34%)	0.01 (54%)
	NH4-N	0.05	0.09	0.05 (29%)	0.01 (56%)
	Runoff	0.0001	0.07	0.02 (54%)	0.01 (32%)
	Erosion	0.0001	0.11	0.01 (52%)	0.01 (40%)
Future 100 yrs (A2 scenario)	Biomass	0.001	0.001	0.01 (56%)	0.05 (5%)
	Grain yield	0.001	0.001	0.01 (45%)	0.14 (5%)
	Soil carbon	0.0001	0.10	0.001 (62%)	0.05 (23%)
	NO3-N	0.001	0.07	0.02 (48%)	0.05 (31%)
	NH4-N	0.001	0.09	0.05 (32%)	0.02 (46%)
	Runoff	0.0001	0.14	0.01 (64%)	0.05 (30%)
	Erosion	0.0001	0.22	0.02 (45%)	0.05 (27%)

Variances accounted for by soil types in all ecosystem services were highly significant and ranged from 29% for NH_4-N to 74% for soil carbon based on P-scenario and from 32% for NH_4-N to 64% for runoff, based on the F-scenario. Variances accounted for by the interaction between crop rotations and soil types were highly significant, except biomass and grain yield based on P-scenario and grain yield based on F-scenario. In general, these variances were smaller in magnitude as compared with those attributed to differences between soil types with three exceptions: larger variance estimates due to the interaction between crop rotations and soil types were predicted for NO_3-N and NH_4-N based on P-scenario and for NH_4-N based on F-scenario.

A number of ecosystem services are expected to be differentially impacted by future climate change scenario depending on how the soil types may react to changes in future weather variables. More variation in NO_3-N (48 versus 34%) and runoff (64 versus 54%) will be accounted for by future climate change scenario, whereas slightly less variation in biomass (56 versus 61%), soil carbon (62 versus 74%), and erosion (45 versus 52%) will be accounted for by the same scenario. These projected changes may or may not be associated with changes in variances accounted for by the interaction between crop rotations and soil types (Table 2). Variances accounted for by this interaction are expected to increase for soil carbon (12 versus 23%) and to decrease for NO_3-N (54 versus 31%), NH_4-N (56 versus 46%), and erosion (40 versus 27%) in response to future climate change scenario. Soil carbon is a variable of interest in studies involving crop rotations and their interaction with soil types and climate change scenarios. The simulation study suggested that there will be a slight increase (4%) in soil carbon sequestration due to increased CO_2 as projected by future climate change scenario. Also, soil carbon is one of a few variables where soil types and their interaction with crop rotations accounted for large (86 and 85% based on P- and F-scenarios, resp.) portions of its variation.

Response of Soil Types to Climate Change

Each soil type was 100% correctly classified based on its biophysical properties and its reaction to past and future climate change scenarios (Figure 3). Two discriminant functions, with significant contributions of all ecosystem services (i.e., biophysical characteristics) based on their Wilks' λ and R2values (except biomass and grain yield), accounted for 0.89 and 0.11 of variation between soil types, respectively, due to past, and for 0.92 and 0.08 of variation, respectively, between soil types due to future climate change scenarios. Soil carbon, runoff, erosion, NO_3-N, and NH_4-N, in decreasing order, significantly contributed to this discrimination based on their Wilk's λ and $R2$ values (Figure 3). Similarly, runoff, erosion, soil carbon, NO_3-N, and NH_4-N, in decreasing order, significantly contributed to this discrimination based on the future climate change scenario.

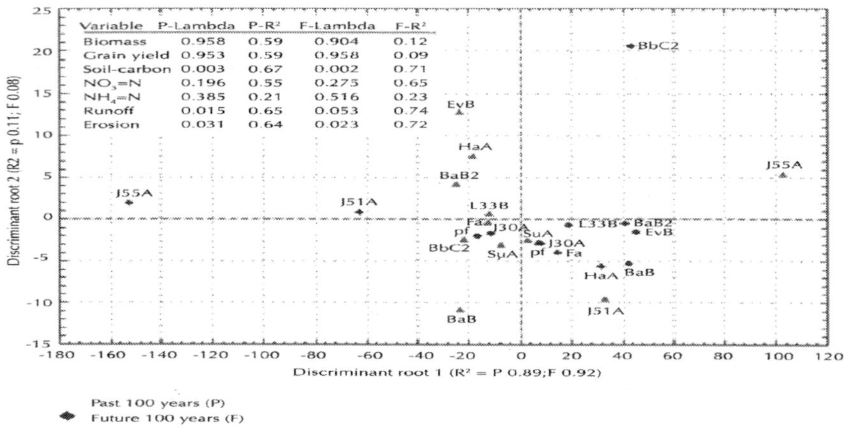

(a)

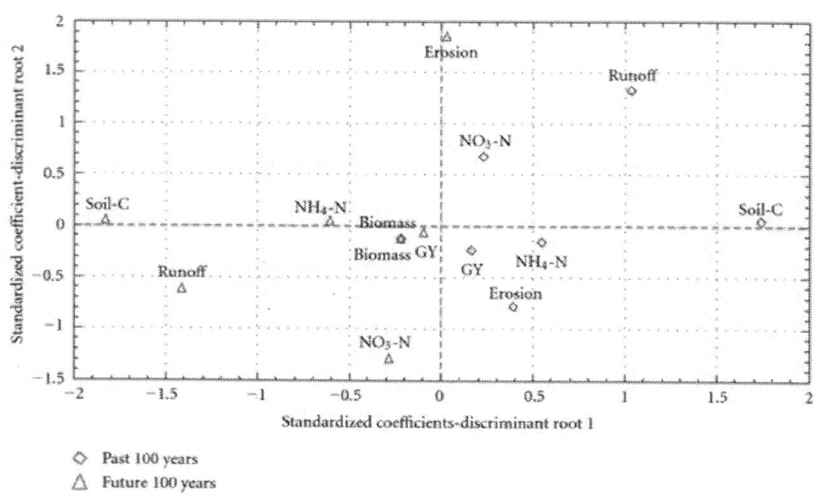

(b)

Figure 3: Scatter plot and loadings of 12 soil types on the first and second discriminant roots, and test statistics (inset: P- and F-lambda and *R*2) and the standardized coefficients of seven ecosystem services as impacted by past (P-) and future (F-) climate change in the Chippewa River Watershed, MN.

The scatter of soil types along the first discriminant root (which accounted for most variation) indicated that a few of these soil types (Fa, J30A, L33B, Pf, and SuA) may experience relatively small variation in their biophysical properties due to future climate change. The remaining soil types may undergo medium (BaB, BaB2, EvB, and HaA) or large (BbC2, J51A, and J55A) variations. These future shifts in soils' biophysical characteristics are largely quantified by their standardized coefficients on both discriminant roots (Figure 3(b)). Soil carbon and runoff had larger standardized coefficients as compared to erosion, NO_3-N, and NH_4-N, while biomass and grain yield had the smallest standardized coefficient values.

Spatiotemporal Variation in Ecosystem Services

Correlation coefficients between past and future normalized values, and regression models (Figure 4) describing the nonlinear temporal changes in ecosystem services, indicated that soil types (representing spatial ecosystem services), most likely, will respond to climate change in different manners. The relatively small NH_4-N data, as compared to NO_3-N data, was not included in Figure 4 for convenience. Temporally, past and future ecosystem services were significantly correlated. The largest r-values were found for soil carbon ($r = 0.98$; $P < 0.001$) and erosion ($r = 0.97$; $P < 0.001$); intermediate values were found for biomass ($r = 0.78$; $P < 0.01$), grain yield ($r = 0.76$; $P < 0.01$), and NO_3-N ($r = 0.73$; $P < 0.01$), and smallest value was found for runoff ($r = 0.56$; $P < 0.05$); however, the variations explained by the nonlinear regression models expressing future values of ecosystem services as function of past values were all significant ($P < 0.05$) and ranged from small ($R^2 = 0.31$ for runoff) to large ($R^2 = 0.96$ for soil carbon).

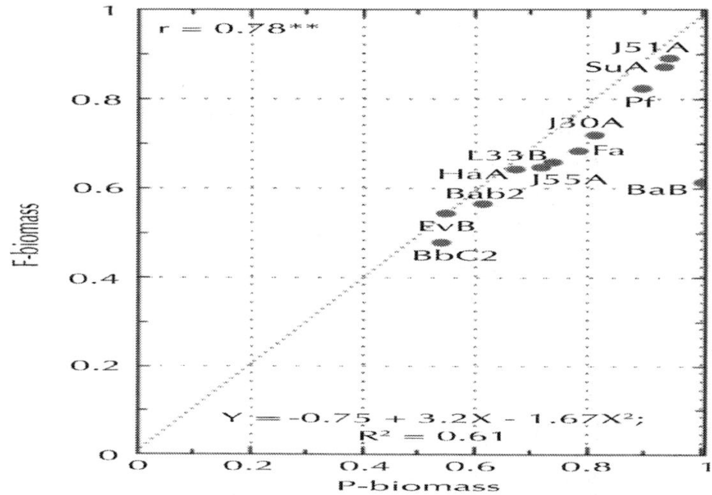
r = 0.78**
J51A
SuA
Pf
J30A
Fa
L33B
HaA
J55A
BaB
Bab2
EvB
BbC2
Y = -0.75 + 3.2X - 1.67X²;
R² = 0.61
F-biomass
P-biomass
0
0.2
0.4
0.6
0.8
1

(a)

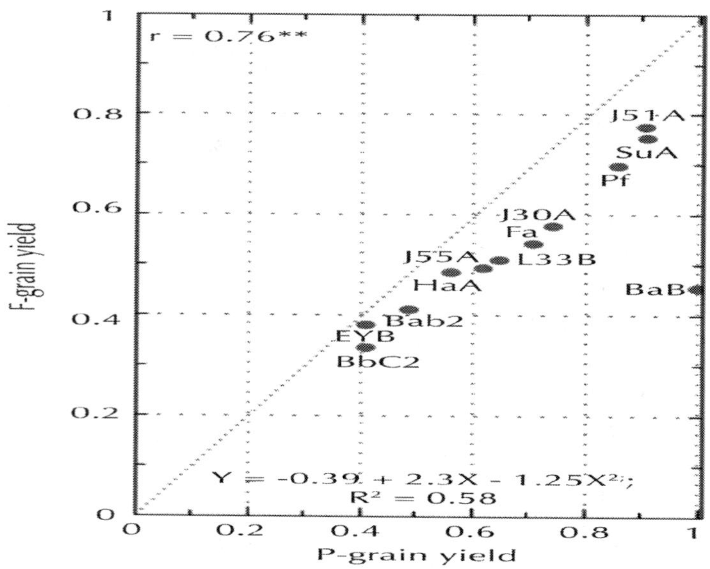
r = 0.76**
J51A
SuA
Pf
J30A
Fa
J55A
L33B
HaA
BaB
Bab2
EYB
BbC2
Y = -0.39 + 2.3X - 1.25X²;
R² = 0.58
F-grain yield
P-grain yield
0
0.2
0.4
0.6
0.8
1

(b)

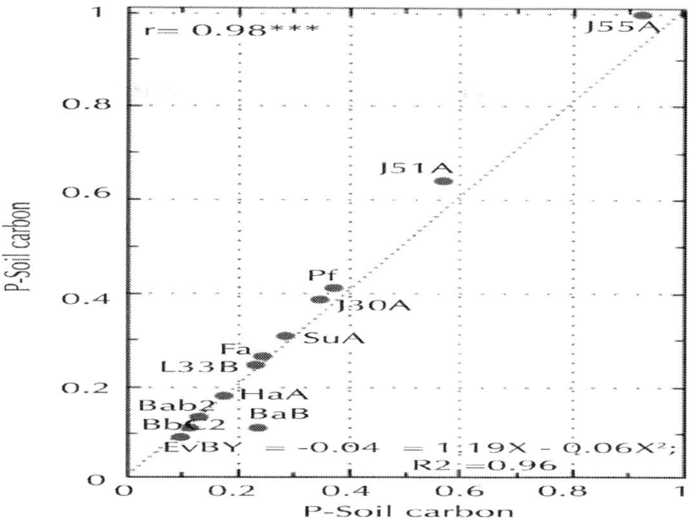
r= 0.98***
J55A
J51A
Pf
J30A
SuA
Fa
L33B
HaA
Bab2
BaB
BbC2
EvBY = -0.04 = 1.19X - 0.06X²;
R2 =0.96
0
0.2
0.4
0.6
0.8
1
P-Soil carbon
P-Soil carbon

(c)

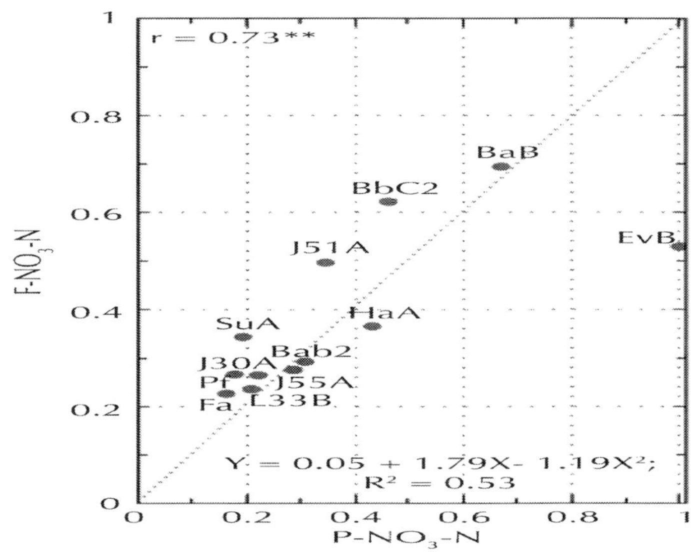
r = 0.73**
BaB
BbC2
EvB
J51A
SuA
HaA
Bab2
J30A
Pf
J55A
Fa
L33B
Y = 0.05 + 1.79X- 1.19X²;
R² = 0.53
0
0.2
0.4
0.6
0.8
1
F-NO3-N
P-NO3-N

(d)

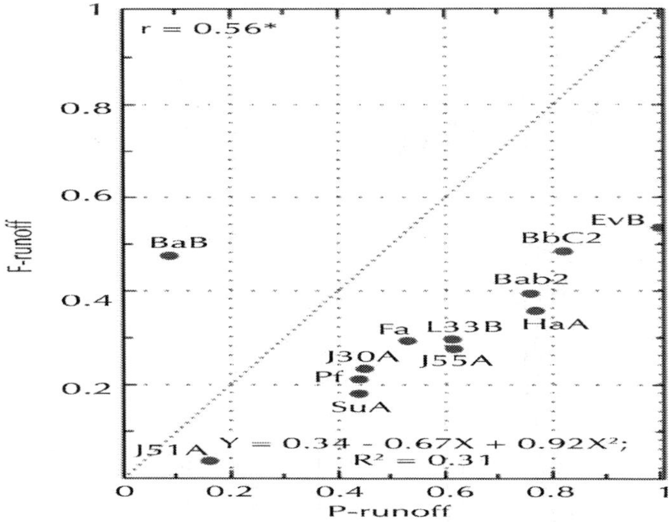

(e)

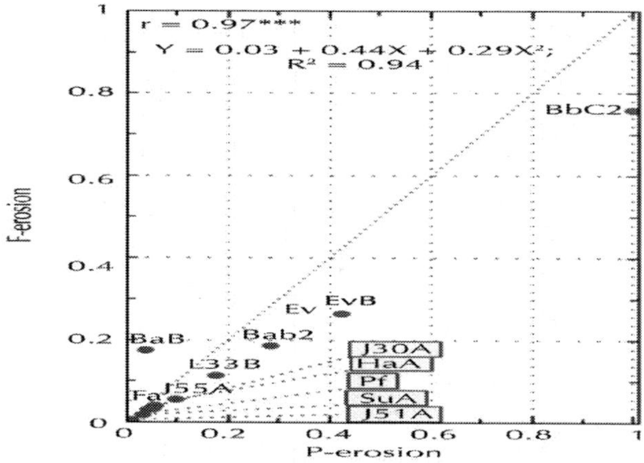

(f)

Figure 4: Prediction models and their test statistics of future (F-) levels of eco-system services as functions of their respective past (P-) levels and based on response of 12 soil types to past (P-) and future (F-) climate change scenarios in the Chippewa River Watershed, MN.

Regression models (expressing future ecosystem services as functions of their respective past ones; Figure 4) for biomass and grain yield mirrored each other as to statistical descriptors, most productive (BaB; P-biomass and P-grain y i e l d = 1; F-biomass and F-grain yield = ~0.6 and ~0.42, resp., Figure 4) and least productive (BbC2) soil types, and temporal changes in productivity. Biomass and grain yield under future climate change scenario are predicted to decline, with largest expected values being 0.92 and 0.78 of their respective maximum values. The soil type BaB would be the most productive under past (P-biomass and P-grain yield) but not under future climate change scenario where J51A would be the most productive soil type. If we consider the 0.5 as the agronomically and environmentally acceptable cutoff level for all ecosystem services, then BbC2 would be the only soil type to fall below the biomass cutoff level, and along with EvB and Bab2 soil types would fall below the grain yield cutoff level under future climate change scenario.

Carbon sequestration is expected to increase slightly under future climate change scenario, with two soil types (J55A and J51A) having values above 0.5 while the remaining 10 soil types falling below the 0.5 cutoff level. Runoff and soil erosion are expected to decrease and NO_3-N leaching is expected to increase under future climate change scenario. The EvB soil type would experience the largest NO_3-N leaching and runoff, and BbC2 would experience the largest soil erosion under past climate change scenario. Two soil types (BbC2 and BaB) would experience the largest NO_3-N leaching under future climate change scenario; however, their future values are expected to reach 0.3–0.4 of the maximum.

Most of the nonlinear regression models describing the relationships between each of the six ecosystem services (Figure 4) had negative intercepts and a negative polynomial regression coefficient (except those for runoff and erosion). Notwithstanding the R^2 values (range from 0.73 to 0.98), this suggests that the future levels of biomass, grain yield, soil carbon, and NO_3-N leaching under future climate change scenario would reach a maximum then decrease in relation to their past values, whereas runoff and erosion are expected to decrease with low (R^2= 0. 3 1) and high (R^2= 0. 9 7) probability, respectively, under future climate change scenario. The extreme erosion (and to some extent runoff) values for the BbC2 soil type (slope of 6–12%) may have skewed the distribution of the normalized values of these two ecosystem services.

Perennial Crop Effect on Ecosystem Services

The effect of increased proportion of a perennial crop (i.e., alfalfa as a nitrogen-fixing forage legume crop) on ecosystem services (Figure 5), whether under past or future climate change scenarios, was nonlinear and positive for biomass, grain yield, and carbon sequestration and nonlinear and negative for NO_3-N leaching, runoff, and soil erosion. Maximum expected biomass and grain yield due to the perennial effect would be achieved at 0.45 perennial proportion of the crop rotation under both climate change scenarios. However, the maximum expected biomass and grain yield under future climate change scenario would be 0.97 and 0.8 of their respective maximum values expected under past climate change scenario. In both cases, the conventional corn-soybean crop rotation is expected to produce the least biomass and grain yield under both scenarios. The predictive models estimated biomass under past and future climate change scenarios with R^2 of 0.58 and 0.74, respectively. The respective R^2 values for grain yield were 0.62 and 0.64. The effect of percent perennial in the crop rotation on carbon sequestration was almost linear under both climate change scenarios (regression coefficients in the nonlinear portion of the regression models were 0.02 and 0.03 and were associated with large R^2values). The continuous perennial is expected to sequester the maximum amount of carbon while the corn-soybean crop rotation is expected to sequester the minimum; however, the differences were minimal as compared with other ecosystem services.

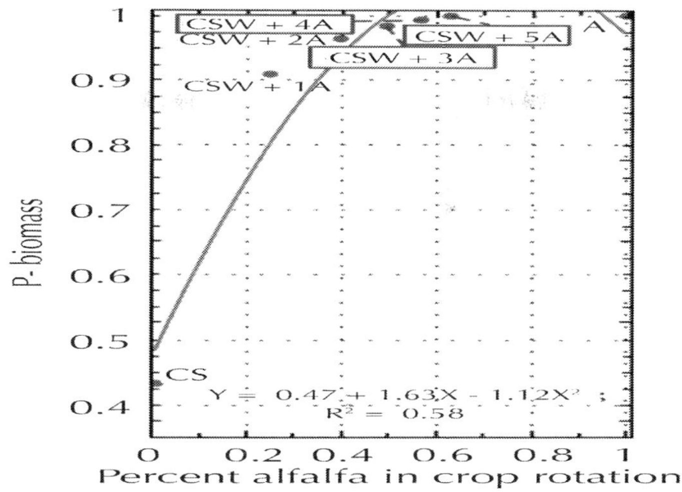
CSW + 4A
CSW + 2A
CSW + 5A
A
CSW + 3A
CSW + 1A
CS
Y = 0.47 + 1.63X - 1.12X² ;
R² = 0.58
P- biomass
Percent alfalfa in crop rotation

(a)

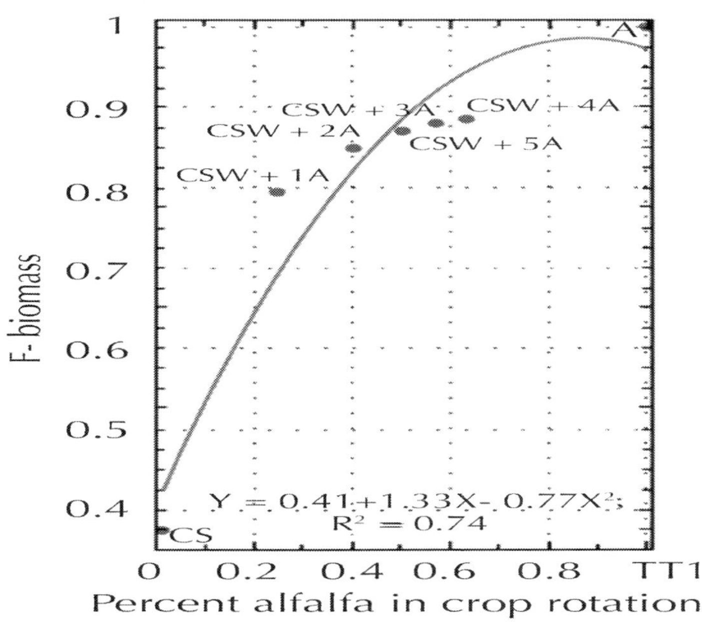
A
CSW + 3A
CSW + 4A
CSW + 2A
CSW + 5A
CSW + 1A
Y = 0.41+1.33X- 0.77X²;
R² = 0.74
CS
F- biomass
Percent alfalfa in crop rotation

(b)

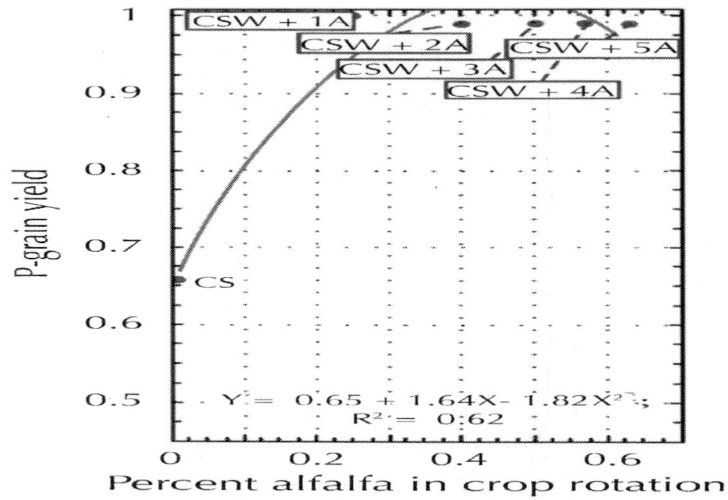
CSW + 1A
CSW + 2A
CSW + 3A
CSW + 4A
CSW + 5A
CS
Y = 0.65 + 1.64X- 1.82X²;
R² = 0.62
P-grain yield
1
0.9
0.8
0.7
0.6
0.5
0
0.2
0.4
0.6
Percent alfalfa in crop rotation

(c)

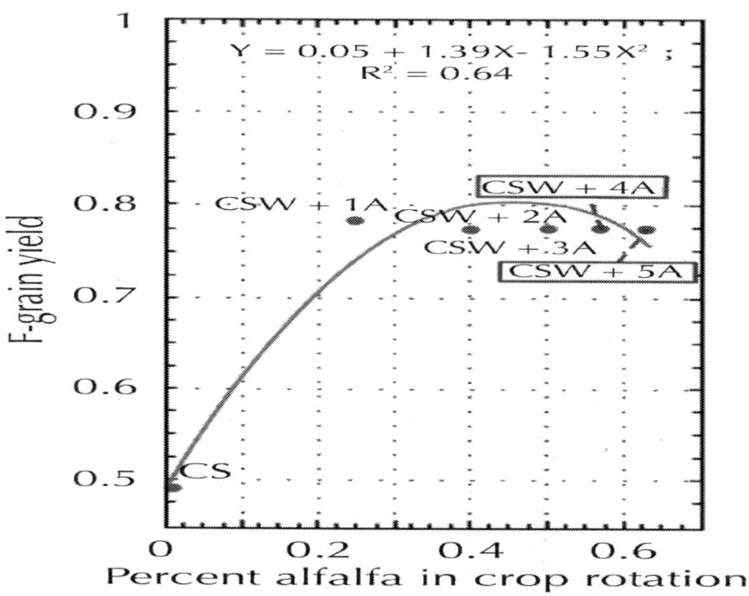
Y = 0.05 + 1.39X- 1.55X² ;
R² = 0.64
CSW + 1A
CSW + 2A
CSW + 3A
CSW + 4A
CSW + 5A
CS
F-grain yield
1
0.9
0.8
0.7
0.6
0.5
0
0.2
0.4
0.6
Percent alfalfa in crop rotation

(d)

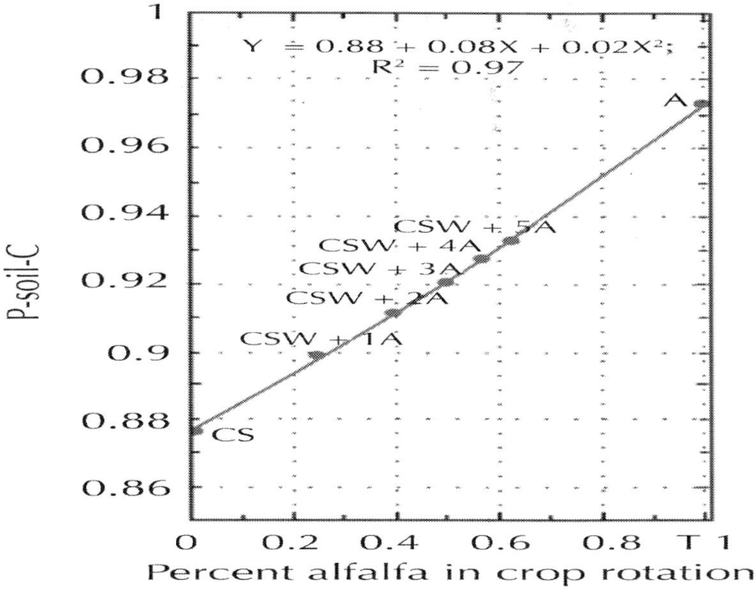
Y = 0.88 + 0.08X + 0.02X²;
R² = 0.97
P-soil-C
Percent alfalfa in crop rotation
CS
CSW + 1A
CSW + 2A
CSW + 3A
CSW + 4A
CSW + 5A
A

(e)

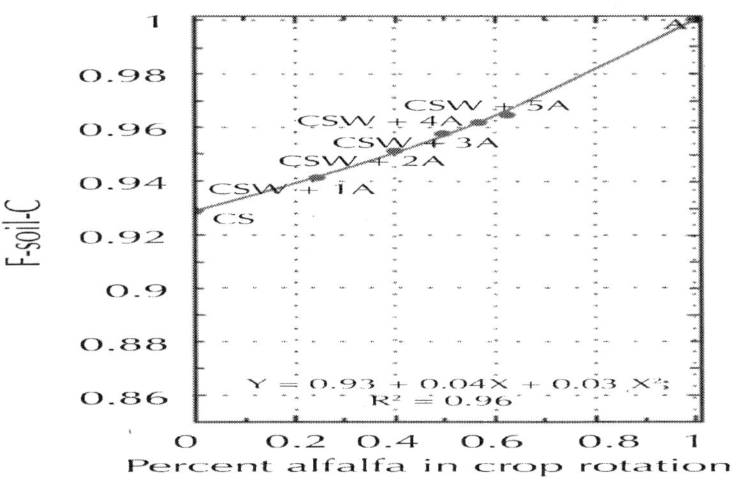
F-soil-C
Percent alfalfa in crop rotation
CS
CSW + 1A
CSW + 2A
CSW + 3A
CSW + 4A
CSW + 5A
A
Y = 0.93 + 0.04X + 0.03 X²;
R² = 0.96

(f)

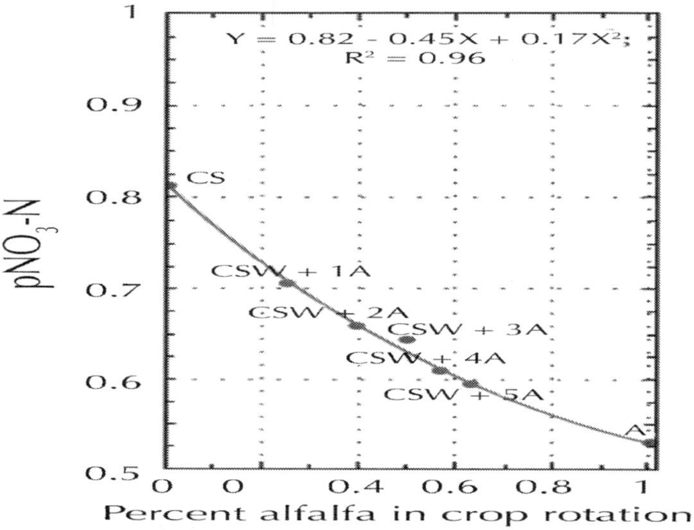
Y = 0.82 - 0.45X + 0.17X²;
R² = 0.96
pNO3-N
CS
CSW + 1A
CSW + 2A
CSW + 3A
CSW + 4A
CSW + 5A
A
Percent alfalfa in crop rotation

(g)

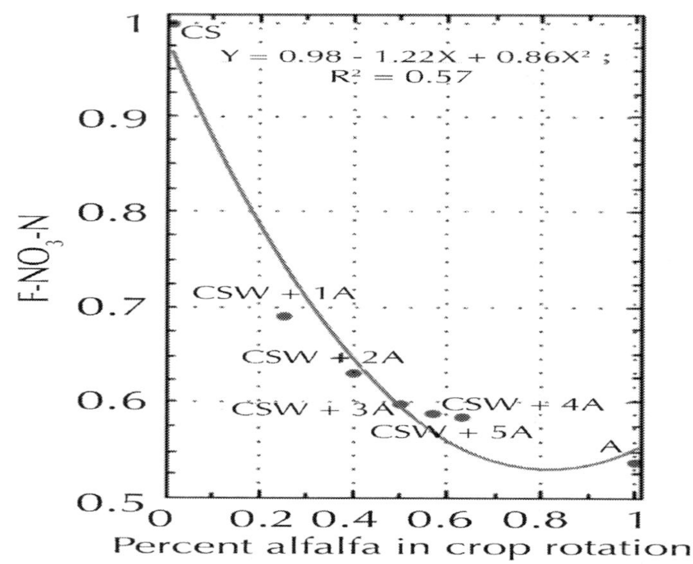
Y = 0.98 - 1.22X + 0.86X² ;
R² = 0.57
F-NO3-N
CS
CSW + 1A
CSW + 2A
CSW + 3A
CSW + 4A
CSW + 5A
A
Percent alfalfa in crop rotation

(h)

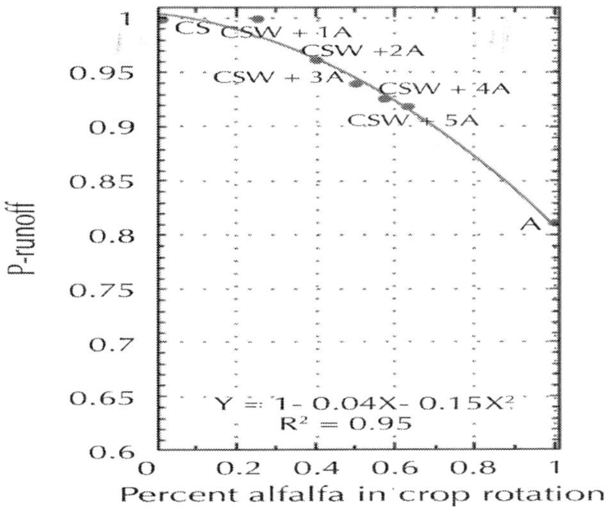
P-runoff
CS
CSW + 1A
CSW +2A
CSW + 3A
CSW + 4A
CSW + 5A
A
Y = 1- 0.04X- 0.15X²
R² = 0.95
Percent alfalfa in crop rotation

(i)

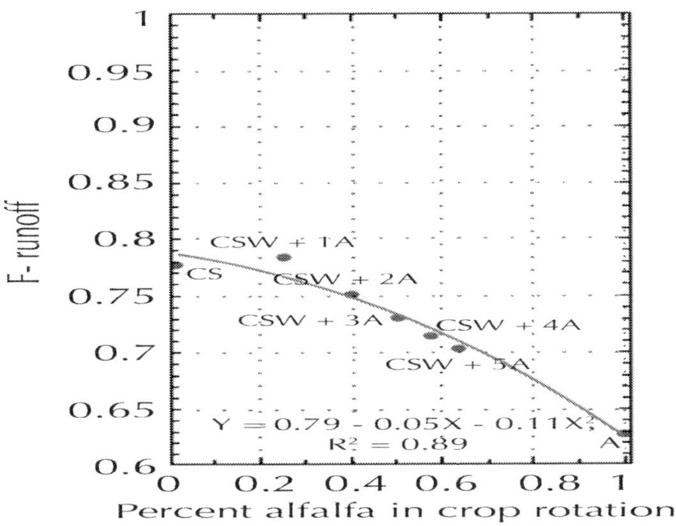
F- runoff
CSW + 1A
CS
CSW + 2A
CSW + 3A
CSW + 4A
CSW + 5A
A
Y = 0.79 - 0.05X - 0.11X²
R² = 0.89
Percent alfalfa in crop rotation

(j)

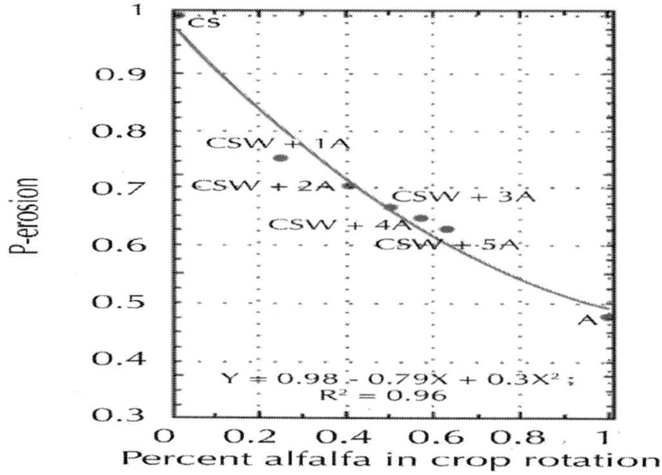

(k)

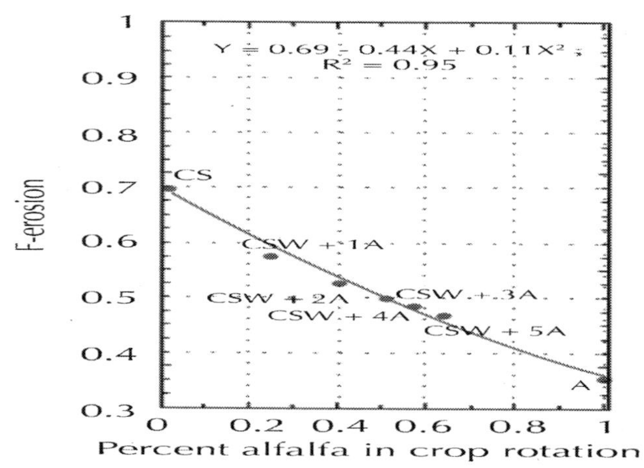

(l)

Figure 5: Prediction models and their test statistics of ecosystem services as functions of percent perennial (alfalfa) in crop rotations and based on past (P-) and future (F-) climate change scenarios in the Chippewa River Watershed, MN.

The level of NO_3-N leaching is expected to decline under future compared with past climate change scenario as quantified by their respective linear regression coefficients (b 1 = − 1. 2 2 and −0.45), nevertheless, the corn-soybean crop rotation would experience the maximum NO_3-N leaching level. The regression models predicted a smaller decrease in NO_3-N leaching (nonlinear regression coefficient, b 2 = 0. 1 7) under past, and a larger decrease (b 2 = 0. 8 6) under future, climate change scenario when the perennial crop occupies more than 0.8 of the crop rotation. Runoff and soil erosion predictions followed trends similar to that of NO_3-N leaching. The corn-soybean crop rotation had the largest runoff and soil erosion predicted values under past and future climate change scenarios. Runoff is predicted to amount to ~0.8 and 0.35 and soil erosion 0.5 and 0.35 of their respective maximum values under continuous perennial due under past and future climate change scenarios, respectively. The nonlinear regression coefficients in both cases (b 2 = − 0. 1 1 and −0.15, resp.) were larger than their respective linear regression coefficients (b 1 = 0. 0 4 and −0.05, resp.). Nonetheless, the percent perennial in the crop rotation predicted these ecosystem services with large probabilities (R^2> 0. 8 9).

Predictability of Future Ecosystem Services under Climate Change

Partial least squares regression modeling (PLS; Figure 6) captured maximum variation in independent (P-) and dependent (F-climate change scenario) ecosystem services at the prediction phase (R^2X and R^2Y, resp.), as well as in the dependent ecosystem services, at the validation phase (Q^2Y) of model development. The PLS regression models differed as to all three statistics (i.e., R^2X, R^2Y, and Q^2Y). These differences can be attributed to differences between physical and chemical characteristics of different soil types and to the nature and strength of association between different ecosystem services within a particular soil type.

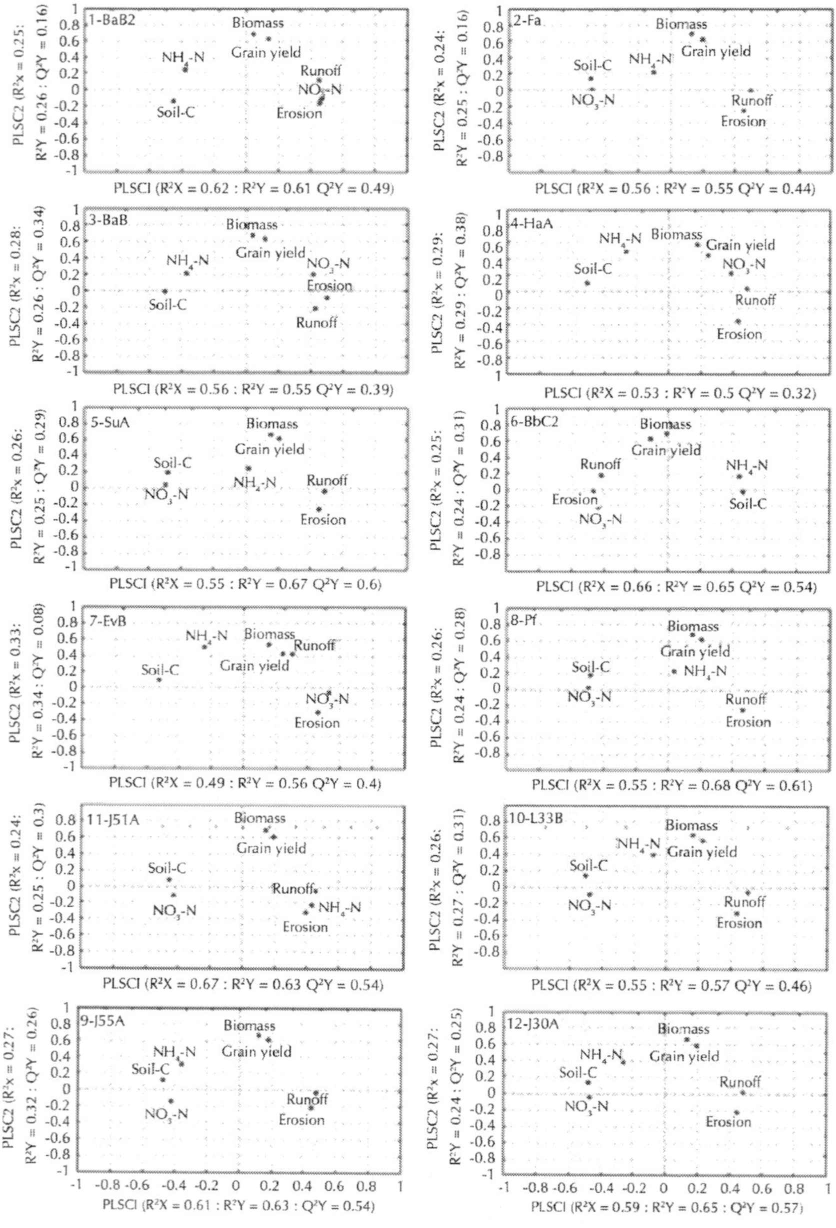

Figure 6: Prediction and validation models and their test statistics of future ecosystem services as functions of past ecosystem services in 12 soil types in the Chippewa River Watershed.

Two partial least squares regression components (PLSC1 and PLSC2) accounted for a maximum (Q^2Y = 0. 8 9) to a minimum (Q^2Y = 0. 4 8) of the total variation in predicting future (F-) as a function of past (P-) ecosystem services. However, the portion accounted for by each component differed among soil types. Loadings (i.e., correlation coefficients between an ecosystem service and the PLSC) primarily on PLSC1, and to a lesser extent on PLSC2, quantify the power and direction (positive or negative effect) of that ecosystem service in predicting future ecosystem services. Biomass and grain yield loadings on PLSC1 were positive but small as compared with the rest of ecosystem services, except in BbC2 where their loadings on PLSC1 were small and negative. The remaining ecosystem services were mostly clustered in tow groups with positive (e.g., NO_3-N, runoff, and erosion) and negative (e.g., soil carbon and NH_4-N) loadings on PLSC1, with some variation in these groupings depending on soil type. Loadings on PLSC2 were generally smaller than loadings on PLSC1.

The carbon-nitrogen associations with, and loadings on, PLSC1 and PLSC2 separated 11 of the 12 soil types into three groups. The BbC2, unlike other soil types, displayed positive loadings of soil carbon and NH_4-N and negative loadings of NO_3-N on PLSC1. Soil carbon and NH_4-N had negative and NO_3-N had positive loadings on PLSC1 in four soil types (Bab2, BaB, EvB, and HaA) regardless of their loadings on PLSC2. Three soil types (SuA, J55A, and Pf) had negative loadings of soil carbon and NO_3-N and positive loading of NH_4-N on PLSC1 regardless of their loadings on PLSC2. The remaining four soil types (J51A, Fa, L33B, and J30A) had negative loadings of all three ecosystem services on PLSC1, and mostly negative loadings of NO_3-N on PLSC2. The NO_3-N was associated with runoff and erosion with positive loadings on PLSC1 in four soil types (Bab2, BaB, EvB, and HaA) and negative loadings in one soil type (BabC2). On the other hand, NH_4-N was associated with soil erosion and runoff with positive loadings on PLSC1in three soil types (SuA, J55A, and Pf). Finally, runoff and soil erosion loaded positively, and NO_3-N and NH_4-N loaded negatively on PLSC1 and were associated with soil carbon in the remaining four soil types (J51A, Fa, L33B, and J30A).

DISCUSSION

Agricultural activities influence the environment and many agro-ecosystem services through their effect on water quality, nutrient cycling, microclimate control, biodiversity conservation, and carbon sequestration, at local, regional, and global scales [30]. Pressing challenges of climate change, natural resource depletion, and environmental degradation triggered renewed interest in current and future status and value of ecosystem services by scientists, farmers, and policy makers at a global [31], regional [32], and local levels, including the CRW [6, 19]. Agro ecosystems in the CRW are more complex than other resource systems due to the large number of manipulated biophysical processes, inputs, outputs, and their interactions [6, 19, and 31]. Productivity of these agro ecosystems is heavily constrained by physical ecosystem components, especially climate (e.g., rainfall and temperature) and soils; therefore, the main focus in this analysis was more on relationships among components and processes than on single-factor cause-effect relationships.

The CRW is a source of provisioning (food, feed and biofuel), regulating (climate regulation, flood regulation, and water quality control), cultural (recreational and aesthetic values), and supporting services (nutrient cycling) [6]. Historical land-use changes have had a significant effect on the hydrology of the watershed and therefore on the provision of these ecosystem services. Due to changing agricultural demands, land use in the watershed will continue to change, and we should be able to derive insight into more sustainable management practices for the future [16]. The environmental implications of intensive corn and soybean production in the CRW are serious cause of concern. Runoff and tile drainage from intensively managed fields are well-documented causes of nonpoint source contamination of surface and underground water bodies with sediments and nutrients [26]. Therefore, farmers are facing growing demands to reduce environmental pressure and the greater implementation of diversified agro ecosystems may be a productive strategy to build resilience into future agro ecosystems.

Past and Future Climate

Agricultural production in the Midwest is critically dependent on the weather, especially rainfall amount and distribution during the growing season [17]. We selected the A2 scenario because it represents a very heterogeneous world where economic development is regionally oriented and economic growth and technological change are relatively slow [17, 20]. By mid- and late 21st century, A2 emissions are expected to reach 550 ppm and to exceed those of the A1F1 "high" scenario, respectively [17]. However, despite some similarities between weather variables of past and future weather, future weather based on A2 (business as usual; [17]) is expected to radically depart from the long-term distribution of rainfall, temperature, and CO_2 concentration as evidenced by its statistical moments. Deviations of future temperature (d-Temp), rainfall (d-Rain), and CO_2 (d-CO_2) from past records, being positively correlated with future values of their respective variables, gave insight into the expected reaction of crop rotations and soil types to future climate and into their expected effects on ecosystem services. The amount of variation in several ecosystem services accounted for by differences between past and future climate change scenarios (Tables 1 and 2) are indications of the direction and level of impact future weather may have on these services.

The 100-year period was chosen to allow for the dynamics of the ecological effects of land-use change to play out and be captured in simulation results [33]. Empirical results [13] indicated that trends in rainfall and temperature may explain 35–40% of crop yields under the Corn Belt conditions, whereas, according to projections based on future climate change scenario, for each additional degree (°C) of future warming during summer months, corn and soybean yield could potentially decrease by 13 and 16%, respectively. Nevertheless, future increases in CO_2 may not be effective at boosting productivity of all crops in the CRW; the expected rise in temperature may offset the positive impact of higher CO_2, especially on small grain crops [34, 35].

Effect of Climate, Soil Type, and Their Interaction on Ecosystem Services

All ecosystem services in the CRW are likely to be impacted by climate, soil types, and their interaction (Table 1). However, proper management of the intensively cropped agro ecosystems in CRW under future climate change is expected to improve soil quality and sustain future productivity [32, 36, 37]. In their comprehensive review of the effect of climate change on agriculture, Hatfield et al. [18] concluded that American agriculture, especially in the Midwest, is vulnerable to climate change but has considerable potential for adaptability. In the next few decades, agriculture will have to undergo significant transformation in order to adapt to climate change and protect natural resources and the environment. Therefore, if the current management practices of high external inputs and conventional tillage continued well into the 21st century, ecosystem services, depending on soil type [38], may respond mostly negatively to future climate change, although this response could be mediated by the soil carbon dynamics [38, 39] and modified by appropriate management practices [40, 41]. The latter may include adoption of more complex crop rotations to match soil types [39, 42]. Soil carbon was one of a few variables where soil types and their interaction with crop rotation accounted for large (86 and 85% based on past and future climates, resp. Table 2) portions of its variation. Therefore, matching crop rotations and soil types may constitute the most appropriate mitigation option in the future [43, 44], and adjusting crop sequence proportions may alleviate some of the negative climate impact on ecosystem services [38, 45].

Multivariate Statistical Analyses of Ecosystem Services

Sources of variation in this simulation study were classified as fixed (crop rotation and crops within crop rotations) and random (soil types and the interaction of crop rotations with soil types) (Table 2). The classification made it possible to quantify the variance components that can be attributed to the random factors [29] assuming that the soil types represented a random sample across the landscape in the watershed and they are expected to respond at random to climate change.

Ecosystem services, especially those related to soil processes, are more dependent on crop rotations rather than on crops within rotations (Table 2), provided that a perennial crop is rotated with annuals [39, 46]. The increasing portion of a perennial crop was accredited with most of the beneficial effects of complex crop rotations by positively impacting soil-related processes [47], especially carbon sequestration [48]. However, apart from a slight increase in carbon sequestration in response to future climate change scenario, the simulation study suggested that the reaction of a soil type to different crop rotations will account for most of the variation in potential carbon sequestration under the A2 climate change scenario.

Response of Soil Types to Climate Change

The soil resources in the CRW are typical of intensively cropped soils in the Midwest [37] with a wide range of physiographic, physical, chemical, and biochemical characteristics (Figure 1; [25]). However, the predominant cropping system, which is based on corn-soybean crop rotation and external inputs, is supporting a few ecosystem services, mainly grain production. The scatter diagram of soil types along the first discriminant root (which accounted for most variation in Figure 3) illustrates the wide range of response of soil types to past and future climate change scenarios. A few soil types may not drastically react to future climate change (e.g., SuA, Pf, Fa), while others (e.g., J55A and BbC2) are projected to undergo dramatic shifts in their ecosystem services. Intensive tillage and external inputs for sugar beet crops are typical management practices for J55A, while BbC2 has the steepest position (6–12%) in the landscape and is prone to high erosion. The standardized coefficients, derived from discriminant analysis, parallel their Wilks' lambda and R^2 values and highlight the greater influence of the soil-related processes (i.e., soil carbon, N leaching, runoff, and erosion) in mediating soil-type response to climate change as compared to the smaller response of provisioning service (i.e., biomass and grain yield). The antagonistic relationship between soil carbon and NO_3-N, as quantified by their standardized coefficients on both discriminant functions may indirectly support the generally accepted negative effect of nitrogen fertilizer on net carbon sequestration [49] in some soil types (J55A, BbC2) more than others (e.g., J30A).

The current crop production system, even on a highly productive soil (e.g., BaB and J51A, Figures 3 and4), will have difficulty attaining consistently a goal of achieving acceptable levels of ecosystem services (e.g., <10 mg NO_3-N/L in tile drainage water, even though prudent N fertilization is followed) [50]; therefore, a tradeoff between ecosystem services is inevitable [37]. Nevertheless, long-term experimental results [25, 50] indicated that the type of cropping system and management practices had a greater positive effect on soil performance or ecosystem services than did nitrogen fertilization. Our simulated data, supported by empirical results of a long-term crop rotation experiment within the CRW [25], can be used to prescribe specific management strategies for different soil types or locations in this watershed.

Spatiotemporal Variation in Ecosystem Services

The spatial variation in ecosystem services (Figure 4) is depicted by the scatter of soil types around the 1 : 1 regression line describing the constant (no-change) relationship between the level of these ecosystem services based on past and future climate change scenarios. Apart from biomass and grain yield, soil types are predicted to undergo major non-linear shifts in their position around the 1 : 1 spatial relationship in response to future climate change. The normalized values allowed for direct comparisons to be made between different ecosystem services and would facilitate the development of simple performance-based index to assess multiple agro-ecosystem functions [27]. Soil carbon exhibited the largest spatial shift (a minimum of 0.1 for EvB to maximum of 1.0 for J55A in response to future climate change), whereas the smallest spatial shift was displayed by runoff (a minimum of ~0.05 for J51A to a maximum of 0.55 for EvB).

The temporal variation is quantified by the non-linear relationships between past and future simulated values of all ecosystem services, except soil carbon, in which the polynomial regression coefficient (b 2 = − 0. 0 6) was not significant (Figure 4). When solving each non-linear equation assuming that the average simulated value during the past 100 yrs (i.e., the X variable) equals 1.0, grain yield (1.44) and soil carbon (1.09) under future climate change are expected to exceed their respective values by 0.44 and 0.09 units, respectively, and with the respective medium (R^2= 0. 5 8) and large (R^2= 0. 9 6) probabilities. Biomass, NO_3-N, runoff, and soil erosion under future climate change scenario are expected to be smaller than the maximum value by 0.22, 0.45, 0.41, and 0.24 units, respectively. The expected decrease in simulated biomass can be explained on the basis of increased proportion of the perennial crop in all crop rotations, except the traditional corn-soybean [41]. Increased proportion of the perennial crop in the crop rotation can also explain the expected favorable increase in grain yield, due to the favorable residual nitrogen fertilizer effect [51], and carbon sequestration [41, 47], and decrease in nitrogen leaching, runoff, and soil erosion [52, 53], due to deeper and more vigorous root systems.

The seemingly small increment of carbon sequestered under future climate change scenario can be attributed to the positive feedback in which increases in global temperature (Figure 2) may lower the ability of soils to sequester carbon [54]. In addition, cropping intensification by including more perennials in the crop rotation may result in eventual saturation of the soil profile [48]; hence the small positive increment of carbon sequestered after a 100-year simulation run.

Perennial Crop Effect on Ecosystem Services

Strategies to optimize crop rotations and deliver ecosystem services need to be developed and tested in the CRW. Current simulation results are expected to contribute to this objective and are supported by empirical results of a long-term crop rotation experiment carried out within the CRW [25]. Those results do indicate some potential rotation strategies for different soil types or locations in the watershed. Simulation and empirical results suggested that adding a perennial to the corn-soybean crop rotation can have substantial positive effects on multiple ecosystem services [26, 47]. Growing alfalfa continuously for 8 years may result in larger biomass and sequestered carbon compared to a more complex crop rotation that includes grain crops (e.g., CSW+4A). Including alfalfa to enhance the complexity of a crop rotation was reported to have increased soil organic carbon sequestration by 20 g C $m^{-2}yr^{-1}$ as compared to the corn-soybean rotation [49]. Perennials can mobilize and retain nutrients and water very effectively over extended periods of time as compared with annual crops, thus providing a mechanism to mitigate climate change, and reduce runoff, nutrient loss, and soil erosion [53]. These perennial advantages can be attributed to the fact that more residue diversity and the concomitant increased microbial activity lead to faster decomposition and loss of more carbon from the residue of a complex crop rotation as compared to continuous alfalfa [55].

Regression models suggested that simulated and normalized biomass and grain yield under past and future climate change scenarios are expected to reach their maximum at about 80 and 45% perennial in the crop rotation (Figure 5). Therefore, farmers may have to choose between maximizing biomass or grain yield under a particular management and climate change scenario [32, 40], knowing that a single year of alfalfa is expected to boost grain yield of the succeeding corn crop by ~30% as compared to corn-soybean rotation ([26]; Figure 5). However, this positive effect is expected to gradually diminish as indicated by the negative non-linear regression coefficients for biomass and grain yields ($b\ 2 = -1.1\ 2$ and -1.82, resp.) under past and future climate change scenarios ($b\ 2 = -1.8\ 2$ and -1.55, resp.).

Runoff, soil erosion, and NO_3-N leaching mirrored each other and are expected to decline under future climate change scenario. The most likely factors, besides the perennial crop effect, are future rainfall patterns and temperature regimes and their interactions (Figure 2). However, NO_3-N leaching under future climate change scenario, unlike runoff and soil erosion, was estimated with less certainty (R_2= 0. 5 7). Nitrate leaching from a corn-soybean crop rotation is expected to be the highest and is bound to increase under future climate change scenario (Figure 5). Greater NO_3-N concentrations and drain flows from corn-soybean rotations resulted in annual losses of NO_3-N that were 35 times greater than NO_3-N losses in a perennial system [9]. Other studies [8] estimated annual losses from the traditional corn-soybean rotation at about 30% of applied nitrogen. The perennial crop effect in reducing nitrate leaching is expected to be stronger under future climate change scenario (~50% reduction due to continuous alfalfa) and would level off at about 50–60% perennial in the crop rotation.

Predictability of Future Ecosystem Services under Climate Change

The partial least squares regression modeling procedure used to predict future ecosystem services is more reliable than ordinary regression because it takes into consideration the variation in the dependent variables, the independent variables, and their joint variation [28, 29]. Furthermore, the validation phase of PLS modeling improves the level of prediction by using a rigorous cross-validation procedure [29]. The level of (un)certainty in the validation models, quantified by the *Q2Y* values, accounted for by PLSC1, ranged from 0.32 (HaA) to 0.61 (Pf) and by PLSC2 from 0.08 (EvB) to 0.38 (HaA). Many factors contributed to these results including patterns and levels of association between dependent and independent variables (i.e., ecosystem services) in different soil types. Also, the uncertainty of the climate model [53], especially the fertilization effect of and response of different crops and (future) crop varieties to CO_2, may have contributed to these results.

CONCLUSIONS

Ecosystem services, whether provisioning, regulating, or supporting services, provided by the CRW are being threatened by anthropogenic as well as climatic factors. Although there will be an antagonistic relationship between ecosystem's capacity to simultaneously sustain the availability of provisioning and regulating ecosystem services, complementarities among different types may increase and the strength of their tradeoffs may diminish with the spatial complexity of the agricultural landscape. Therefore, an in-depth understanding of agro ecosystems is essential for the future of land-use change in the CRW. The simulation results offer strategies to optimize site-specific crop rotations in this watershed under projected climate change. The simulated effect of increasing perennial land use in agro ecosystems across the watershed on several biophysical processes suggested that farmers can help improve environmental health through increased perennial cover with sustained carbon sequestration and the concomitant reduction in soil erosion, runoff, and nutrients leaching. Diversifying the corn-soybean crop rotations by including a perennial crop, especially in erosion-prone soil types and locations in the watershed, would provide a strategy to mitigate negative environmental effects caused by corn and soybean production while providing an additional source of income based on new regional markets for food and biomass from perennials and diverse crops. Adaptive management, where stakeholders contribute to optimize resource use and minimize anthropogenic and climatic effect on the production base within the watershed, is expected to help develop multifunctional production systems that can produce standard commodities as well as a wide range of other system services.

ACKNOWLEDGMENTS

The authors thank The Walton Family Foundation for financial support, Jon Starr for his technical assistance, and Beth Burmeister for editing the paper. This study was supported by the National Institute of Food and Agriculture, USDA, under Agreement no. 2010-65615-20630. The use of trade, firm, or corporation names in this publication is for the information and convenience of the reader. Such use does not constitute an official endorsement or approval by the United States Department of Agriculture (USDA) or the Agricultural Research Service (ARS) to the exclusion of others that may be suitable. USDA is an equal provider and employer.

REFERENCES

1. Millennium Ecosystem Assessment: Ecosystems and Human Well-Being: The Assessment Series, vol. 4, Island Press, Washington, DC, USA, 2005.
2. K. E. Schilling, K. S. Chan, H. Liu, and Y. K. Zhang, "Quantifying the effect of land use land cover change on increasing discharge in the Upper Mississippi River," Journal of Hydrology, vol. 387, no. 3-4, pp. 343–345, 2010.
3. E. J. Nelson and G. C. Daily, "Modelling ecosystem services in terrestrial systems," F1000 Biology Reports, vol. 2, no. 1, article 53, 2010.

4. G. Boody, B. Vondracek, D. A. Andow et al., "Multifunctional agriculture in the United States,"BioScience, vol. 55, no. 1, pp. 27–38, 2005.

5. P. Wymar, Chippewa River Watershed Monitoring Summary 2007: Reading the Land in the River. Montevideo, MN. Chippewa River Watershed Project, 2007.

6. N. Jordan, G. Boody, W. Broussard et al., "Sustainable development of the agricultural bio-economy," Science, vol. 316, no. 5831, pp. 1570–1571, 2007.

7. S. Polasky, E. Nelson, D. Pennington, and K. A. Johnson, "The impact of land-use change on ecosystem services, biodiversity and returns to landowners: a case study in the state of Minnesota," Environmental and Resource Economics, vol. 48, no. 2, pp. 219–242, 2011.

8. R. W. Malone, L. Ma, D. L. Karlen et al., "Empirical analysis and prediction of nitrate loading and crop yield for corn-soybean rotations," Geoderma, vol. 140, no. 3, pp. 223–234, 2007.

9. D. R. Huggins, G. W. Randall, and M. P. Russelle, "Subsurface drain losses of water and nitrate following conversion of perennials to row crops," Agronomy Journal, vol. 93, no. 3, pp. 477–486, 2001.

10. S. Secchi, P. W. Gassman, M. Jha, L. Kurkalova, and C. L. Kling, "Potential water quality changes due to corn expansion in the Upper Mississippi River Basin," Ecological Applications, vol. 21, no. 4, pp. 1068–1084, 2011.

11. W. Broussard and R. E. Turner, "A century of changing land-use and water-quality relationships in the continental US," Frontiers in Ecology and the Environment, vol. 7, no. 6, pp. 302–307, 2009.

12. P. Laterra, M. E. Orúe, and G. C. Booman, "Spatial complexity and ecosystem services in rural landscapes," Agriculture, Ecosystems and Environment. In press.

13. C. J. Kucharik and S. P. Serbin, "Impacts of recent climate change on Wisconsin corn and soybean yield trends," Environmental Research Letters, vol. 3, no. 3, Article ID 034003, 2008.

14. C. J. Herringshaw, T. W. Stewart, J. R. Thompson, and P. F. Anderson, "Land use, stream habitat and benthic invertebrate assemblages in a highly altered Iowa watershed," American Midland Naturalist, vol. 165, no. 2, pp. 274–293, 2011.
15. A. D. Nicks, R. D. Williams, and J. R. Williams, "Regional analysis of erosion from agricultural fields using global change scenarios," in Erosion and Sediment Yield: Global and Regional Perspectives, Proceedings of the Exeter Symposium, IAHS Publishers, No. 236, 1996.
16. D. Mao and K. A. Cherkauer, "Impacts of land-use change on hydrologic responses in the Great Lakes region," Journal of Hydrology, vol. 374, no. 1-2, pp. 71–82, 2009.
17. D. J. Wuebbles and K. Hayhoe, "Climate change projections for the United States Midwest,"Mitigation and Adaptation Strategies for Global Change, vol. 9, no. 4, pp. 335–363, 2004.
18. J. Hatfield, K. Boote, P. Fay, et al., "Agriculture," in The Effect of Climate Change on Agriculture, Land Resources, Water Resources, and Biodiversity in the United States, U.S. Climate Change Science Program and the Subcommittee on Global Change Research, Washington, DC, USA, 2008.
19. V. Nangia, P. Wymar, and J. Klang, "Evaluation of a GIS-based watershed modeling approach for sediment transport," International Journal of Agricultural and Biological Engineering, vol. 3, no. 3, pp. 43–53, 2010.
20. Interagency Panel on Climate Change and IPCC, "Impacts, adaptations and vulnerability," inContributions of the Working Group II to the 4th Assessment Report of the Intergovernmental Panel on Climate Change, M. L. Parry, O. F. Canziani, J. P. Palutikof, P. J. van der Linden, and C. E. Hanson, Eds., Cambridge University Press, Cambridge, UK, 2007.
21. A. D. Friend, "Terrestrial plant production and climate change," Journal of Experimental Botany, vol. 61, no. 5, pp. 1293–1309, 2010.
22. A. P. Rayburn and L. A. Schulte, "Landscape change in an agricultural watershed in the U.S. Midwest," Landscape and Urban Planning, vol. 93, no. 2, pp. 132–141, 2009.
23. B. A. Keating, P. S. Carberry, G. L. Hammer et al., "An overview of APSIM, a model designed for farming systems simulation,"

European Journal of Agronomy, vol. 18, no. 3-4, pp. 267–288, 2003.

24. S. K. Mishra and V. P. Singh, Soil Conservation Service Curve Number (SCS-CN) Methodology, Kluwer Academic, Dodrecht, The Netherlands, 2003.
25. A. A. Jaradat and S. L. Weyers, "Statistical modeling of yield and variance instability in conventional and organic cropping systems," Agronomy Journal, vol. 103, no. 3, pp. 673–684, 2011.
26. J. Olmstead and E. C. Brummer, "Benefits and barriers to perennial forage crops in Iowa corn and soybean rotations," Renewable Agriculture and Food Systems, vol. 23, no. 2, pp. 97–107, 2008.
27. M. A. Liebig, G. Varvel, and J. Doran, "A simple performance-based index for assessing multiple agroecosystem functions," Agronomy Journal, vol. 93, no. 2, pp. 313–318, 2001.
28. StatSoft, Inc. STATISTICA (data analysis software system). Version 10, 2011,http://www.statsoft.com/.
29. R. W. Payne, S. A. Harding, and D. A. Murray, The Guide to GenStat Release 10. Part 2: Statistics, VSN International, Hemel Hempstead, UK, 2007.
30. S.T. Lovell, S. DeSantis, C.A. Nathan et al., "Integrating agroecology and landscape multifunctionality in Vermont: an evolving framework to evaluate the design of agroecosystems,"Agricultural Systems, vol. 103, no. 5, pp. 327–341, 2010.
31. F. Caporali, "Agroecology as a transdisciplinary science for a sustainable agriculture," inBiodiversity, Agrofirestry and Conservation Agriculture Reviews, E. Lichtfouse, Ed., vol. 5, Springer Science+Business Media B.V., 2011.
32. A. K. Bhardwaj, P. Jasrotia, S. K. Hamilton, and G. P. Robertson, "Ecological management of intensively cropped agro-ecosystems improves soil quality with sustained productivity,"Agriculture, Ecosystems and Environment, vol. 140, no. 3-4, pp. 419–429, 2011.
33. W. R. Gascoigne, D. Hoag, L. Koontz, B. A. Tangen, T. L. Shaffer, and R. A. Gleason, "Valuing ecosystem and economic services across land-use scenarios in the Prairie Pothole Region of the Dakotas, USA," Ecological Economics, vol. 70, no. 10, pp. 1715–1725, 2011.

34. P. R. Grace, G. Philip Robertson, N. Millar et al., "The contribution of maize cropping in the Midwest USA to global warming: a regional estimate," Agricultural Systems, vol. 104, no. 3, pp. 292–296, 2011.

35. T. O. West, V. Bandaru, C. C. Brandt, A. E. Schuh, and S. M. Ogle, "Regional uptake and release of crop carbon in the United States," Biogeosciences Discussions, vol. 8, no. 1, pp. 631–654, 2011.

36. C. J. Kucharik, "Impact of prairie age and soil order on carbon and nitrogen sequestration," Soil Science Society of America Journal, vol. 71, no. 2, pp. 430–441, 2007.

37. S. S. Snapp, L. E. Gentry, and R. Harwood, "Management intensity—not biodiversity—the driver of ecosystem services in a long-term row crop experiment," Agriculture, Ecosystems and Environment, vol. 138, no. 3-4, pp. 242–248, 2010.

38. C. Caride, G. Piñeiro, and J. M. Paruelo, "How does agricultural management modify ecosystem services in the argentine Pampas? The effects on soil C dynamics," Agriculture, Ecosystems and Environment. In press.

39. A. Meyer-Aurich, A. Weersink, K. Janovicek, and B. Deen, "Cost efficient rotation and tillage options to sequester carbon and mitigate GHG emissions from agriculture in Eastern Canada,"Agriculture, Ecosystems and Environment, vol. 117, no. 2-3, pp. 119–127, 2006.

40. S. W. Choi and B. Sohngen, "The optimal choice of residue management, crop rotations, and cost of carbon sequestration: empirical results in the Midwest US," Climatic Change, vol. 99, no. 1, pp. 279–294, 2010.

41. K. Oorts, P. Garnier, A. Findeling, B. Mary, G. Richard, and B. Nicolardot, "Modeling soil carbon and nitrogen dynamics in no-till and conventional tillage using PASTIS model," Soil Science Society of America Journal, vol. 71, no. 2, pp. 336–346, 2007.

42. G. E. Varvel and W. W. Wilhelm, "Long-term soil organic carbon as affected by tillage and cropping systems," Soil Science Society of America Journal, vol. 74, no. 3, pp. 915–921, 2010.

43. S. Jagadamma, R. Lal, R. G. Hoeft, E. D. Nafziger, and E. A. Adee, "Nitrogen fertilization and cropping system impacts on

soil properties and their relationship to crop yield in the central Corn Belt, USA," Soil and Tillage Research, vol. 98, no. 2, pp. 120–129, 2008.

44. B. Brenda, "Resilience in agriculture through crop diversification: adaptive management for environmental change," BioScience, vol. 61, no. 3, pp. 183–193, 2011.
45. T. H. DeLuca and C. A. Zabinski, "Prairie ecosystems and the carbon problem," Frontiers in Ecology and the Environment, vol. 9, no. 7, pp. 407–413, 2011.
46. H. Blanco-Canqui and R. Lal, "No-tillage and soil-profile carbon sequestration: an on-farm assessment," Soil Science Society of America Journal, vol. 72, no. 3, pp. 693–701, 2008.
47. U. M. Sainju and A. W. Lenssen, "Dryland soil carbon dynamics under alfalfa and durum-forage cropping sequences," Soil and Tillage Research, vol. 113, pp. 30–37, 2011.
48. W. M. Post, R. C. Izaurralde, J. D. Jastrow et al., "Enhancement of carbon sequestration in US soils," BioScience, vol. 54, no. 10, pp. 895–908, 2004.
49. A. E. Russell, D. A. Laird, T. B. Parkin, and A. P. Mallarino, "Impact of nitrogen fertilization and cropping system on carbon sequestration in midwestern mollisols," Soil Science Society of America Journal, vol. 69, no. 2, pp. 413–422, 2005.
50. G. W. Randall, D. R. Huggins, M. P. Russelle, D. J. Fuchs, W. W. Nelson, and J. L. Anderson, "Nitrate losses through subsurface tile drainage in conservation reserve program, alfalfa, and row crop systems," Journal of Environmental Quality, vol. 26, no. 5, pp. 1240–1247, 1997.
51. A. E. Russell, D. A. Laird, and A. P. Mallarino, "Nitrogen fertilization and cropping system impacts on soil quality in midwestern mollisols," Soil Science Society of America Journal, vol. 70, no. 1, pp. 249–255, 2006.
52. H. Asbjornsen, G. Shepherd, M. Helmers, and G. Mora, "Seasonal patterns in depth of water uptake under contrasting annual and perennial systems in the Corn Belt Region of the Midwestern U.S," Plant and Soil, vol. 308, no. 1-2, pp. 69–92, 2008.
53. M. R. O'Neal, M. A. Nearing, R. C. Vining, J. Southworth, and R. A. Pfeifer, "Climate change impacts on soil erosion in Midwest

United States with changes in crop management," Catena, vol. 61, no. 2-3, pp. 165–184, 2005.

54. D. B. Kell, "Breeding crop plants with deep roots: their role in sustainable carbon, nutrient and water sequestration," Annals of Botany, vol. 108, no. 3, pp. 407–418, 2011.

55. Z. Luo, E. Wang, and O. J. Sun, "Can no-tillage stimulate carbon sequestration in agricultural soils? A meta-analysis of paired experiments," Agriculture, Ecosystems and Environment, vol. 139, no. 1-2, pp. 224–231, 2010.

Chapter 5

Ecosystem Services in Agricultural Landscapes: A Spatially Explicit Approach to Support Sustainable Soil Management

Mohsen Forouzangohar[1], Neville D. Crossman[2], Richard J. MacEwan[3], D. Dugal Wallace[4], and Lauren T. Bennett[1]

[1]Department of Forest and Ecosystem Science, The University of Melbourne, 4 Water Street, Creswick, VIC 3363, Australia

[2]Commonwealth Scientific and Industrial Research Organisation (CSIRO) Ecosystem Sciences, PMB 2, Urrbrae, SA 5064, Australia

[3]Future Farming Systems Research Division, Department of Environment & Primary Industries, Bendigo Delivery Centre, VIC 3554, Australia

[4]Agriculture Productivity Group, Department of Environment & Primary Industries, 32 Lincoln Square, North Carlton, VIC 3053, Australia

ABSTRACT

Soil degradation has been associated with a lack of adequate consideration of soil ecosystem services. We demonstrate a broadly applicable method for mapping changes in the supply of two priority soil ecosystem services to support decisions about sustainable land-use configurations. We used a landscape-scale study area of 302 km^2 in northern Victoria, south-eastern Australia, which has been cleared for intensive agriculture. Indicators representing priority soil services (soil carbon sequestration and soil water storage) were quantified and mapped under both a current and a future 25-year land-use scenario (the latter including a greater diversity of land uses and increased perennial crops and irrigation). We combined diverse methods, including soil analysis using mid-infrared spectroscopy, soil biophysical modelling, and geostatistical interpolation. Our analysis suggests that the future land-use scenario would increase the landscape-level supply of both services over 25 years. Soil organic carbon content and water storage to 30 cm depth were predicted to increase by about 11% and 22%, respectively. Our service maps revealed the locations of hotspots, as well as potential trade-offs in service supply under new land-use configurations. The study highlights the need to consider diverse land uses in sustainable management of soil services in changing agricultural landscapes.

INTRODUCTION

That soils are fundamental to a wide range of ecosystem services needs to be acknowledged to avoid further soil degradation and to identify sustainable land-use change. Several recent conceptual works have used an ecosystem service approach to highlight the importance of soils to the sustained prosperity and welfare of humankind [1–4]. Nonetheless, decreases in the supply of soil ecosystem services like water quality regulation and soil structure stabilization are symptomatic both of ongoing loss of soil natural capital and of ongoing disregard of the full range of soil services in production systems [3, 4].

Better integration of ecosystem service knowledge in soil management systems requires holistic yet straightforward approaches to the quantification and mapping of soil ecosystem service supply

[5]. Rutgers et al. [6] and van Wijnen et al. [7] are recent examples, although these studies did not provide approaches for quantifying changes in multiple soil ecosystem services in response to changing management—a critical knowledge gap recognised by Haines-Young et al. [8].

Full consideration of all soil ecosystem services at landscape to regional scales will rarely be possible, necessitating firstly the identification of priority services and secondly the nomination of soil properties to represent these services. Bennett et al. [9] identified 11 "final" soil ecosystem services that are directly utilised to benefit humans (versus supporting "intermediate" services). Priority services are identified as those that offer the greatest benefit within that landscape context but that also act as likely "surrogates" for the provision of other services [9]. All soil services are essentially aggregates of soil processes, where processes are inputs, losses, transfers, and transformations of material and energy [2, 10]. For example, the intermediate soil service "organic matter cycling" (nomenclature of [9]) is the result of various processes (litter comminution, decomposition, and humification), which are expressed as rates. Full temporal quantification of multiple process rates at landscape scales is not feasible. Instead, soil properties (e.g., soil organic carbon content) can be used to represent the end-point of processes/services (e.g., carbon sequestration) and changes in soil properties used as an indication of change in potential service supply [11].

Accurate prediction of changes in soil services/properties will rely both on the direct measurement of properties and on the use of soil-specific biophysical models. In this context, changes in soil properties are simulated continuously in response to changes in soil processes as a result of interactions among soil, water, vegetation, and the atmosphere [12]. There are many soil simulation models, some of which have been recommended for soil services assessment [13]. For example, Aitkenhead et al. [14] introduced the MOSES model for quantitative assessment of selected ecosystem services calibrated for a range of soils in the UK. In Australia, Agricultural Production System Simulator (APSIM) was developed as a modelling framework, consisting of soil, plant, and management component modules, to simulate diverse production systems specific to Australian soils and conditions [15, 16]. APSIM has considerable potential as a tool for modelling changes in soil properties [13] because soil properties are simulated continuously in response to changes in weather, management, and vegetation [15, 16]. It has been applied to the prediction of changes in soil organic carbon (SOC) content as a result of changes in land management [17–19]. In addition, APSIM has been successfully tested in the simulation of soil water balance [20–22].

Soil simulation models are traditionally point-based, whereas soil ecosystem services are best represented as continuous surfaces for decision-making. Interpolation is required and kriging offers an objective and rigorous approach to interpolating between spatially-distributed data points and, thus, to mapping soil services [23, 24]. The use of kriging [25] to produce soil property maps has been demonstrated frequently for SOC [26–29]. Although many studies applied these techniques at field scales with high sample densities [26, 28], some dealt with sparser sampling schemes at much broader scales. For example, Kumar and Lal [27] estimated the spatial distribution of SOC across an area of 117599 km^2 using 920 sample points (~ one sample per 128 km^2). Similarly, Zhang et al. [29] produced a map of SOC for the Republic of Ireland based on kriging interpolations using an average sample density of two per 100 km^2. Clearly, in the context of broad-scale changes in soil properties, a key requirement is to choose a sampling intensity that accommodates cost and logistical constraints, but that is also sufficient for capturing change trends at appropriate scales within accepted levels of error. Here, while demonstrating the advantages of kriging from grid samples for regional estimations of soil properties, McBratney and Webster [24] concluded that the best sampling strategy would be a regular grid scheme with the highest affordable number of observations.

We used a novel combination of methods including field sampling, biophysical modelling (APSIM), and spatial interpolations (e.g., kriging) to produce landscape-scale maps indicating changes in the supply of soil services from current to future management scenarios. We selected a study area in northern Victoria, south-eastern Australia, to illustrate the utility of this approach for considering priority soil services in decisions relating to changing configurations of land-use practices at landscape scales. We focused on two priority services, namely, carbon sequestration and water storage. We used a landholder defined scenario of future land-use configurations (involving increased complexity/diversity) to predict effects of land-use change on the spatial distribution of two soil properties (hereafter "indicators"), namely, SOC content (indicating carbon sequestration) and volume of water stored in soil (indicating water storage). Our aims were to quantify and map the distribution of service indicators (representing priority services) under both current and future land-use configurations and to demonstrate the utility of our approach in supporting land-use decisions by highlighting locations where management resources could be targeted to maintain or enhance the supply of priority soil services [11, 30–32].

METHODS

Approach Overview

Our unique approach (Figure 1) included representative sampling of soils across the study area according to a geostatistically-valid design, analysis of selected soil properties in a subset of reference soil samples, development of mid-infrared calibration models for the prediction of soil properties in the remaining samples, parameterising and configuring the APSIM model to predict indicators (representing priority services) under current and future land-use scenarios, and applying spatial interpolations (kriging) in a GIS environment to produce maps of current and of likely future changes in the distribution of these indicators across the entire study area. In addition, a hotspot map of soil services change was produced to help evaluate the likely trade-offs and opportunities arising from future land-use configurations.

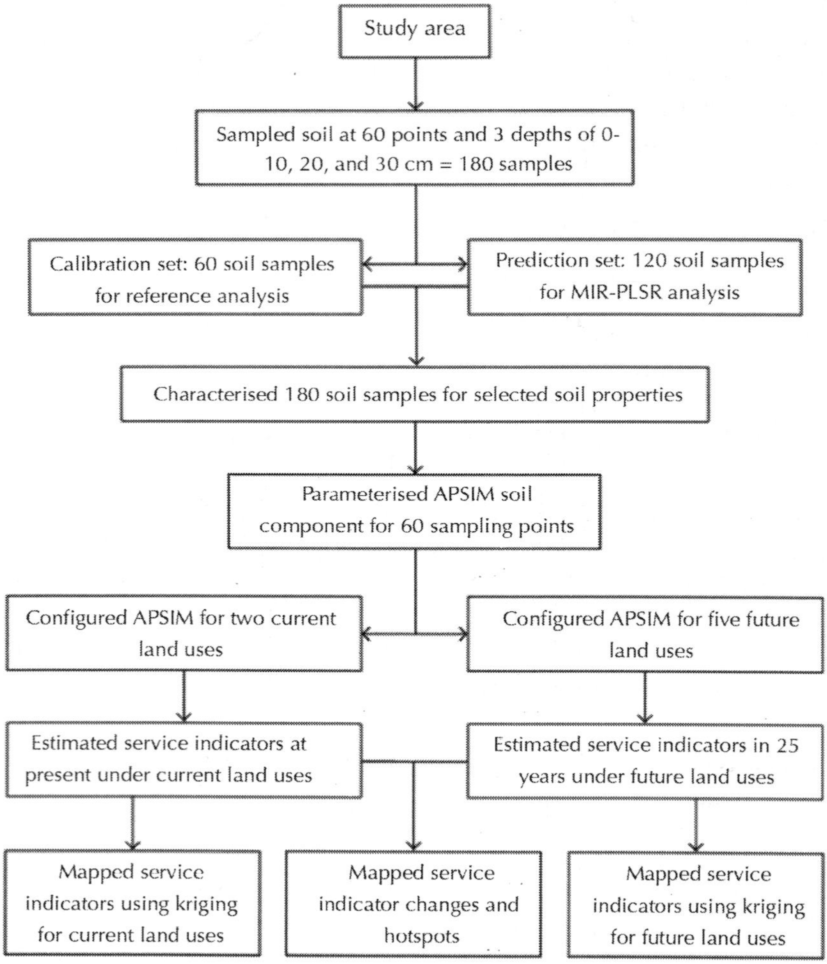

Figure 1: Approach overview, summarising the multiple steps leading to the production of indicator maps (representing priority services) under current and future land-use configurations.

Study Area

This study encompasses a case-study landscape of 30,200 ha (302 km^2) in northern Victoria, between 35° 24.53′ and 35° 41.23′S and 143° 37.72′ and 143° 54.22′E, south-eastern Australia (Figure 2). The region has a warm grassland climate (Australian Bureau of Meteorology—http://www.bom.gov.au/iwk/climate_zones/) and typically experiences hot and dry summers and cold winters and an average annual rainfall of 370mm falling mostly in winter and spring (June to November). Mean monthly minimum temperatures range from 3.3°C (August) to 18.4°C (January), and mean monthly maximum temperatures range from 13.1°C (July) to 35.8°C (January; climate data based on 25-year data from nearby Swan Hill and Kerang weather stations).

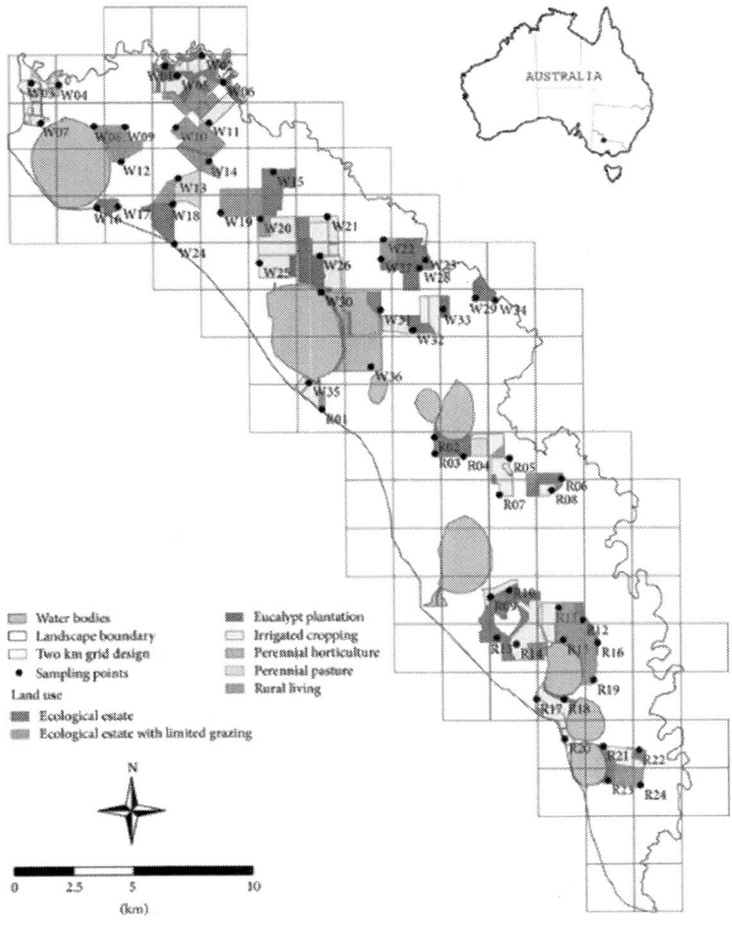

Figure 2: Study area location, the distribution of proposed land uses in the future (25-year) land-use scenario (covering a total area of 6,441 ha) and the associated irregular-grid used to sample soils in this study.

There is no detailed soil survey of the study area. According to the Australian Soil Classification [33] and the Victorian GMU250 geomorphology database [34], the dominant soil orders are Sodosols (or Solonetz in World Reference Base (WRB) soil classification) and Vertosols (or Vertisols in WRB soil classification), generally in the western and eastern sides of the study area, respectively. The region contains several water bodies, some of which are recognised as part of the Ramsar Convention on Wetlands, [35]. The Victorian Land Use Information System 2009 dataset [36] lists the principal land uses within the study area as domestic livestock grazing, mixed farming and grazing, and cereal dry cropping. The landscape has been extensively cleared for agriculture, and the remaining native vegetation is highly fragmented and/or restricted to riparian zones. Historical clearing and intensive agriculture in this landscape have contributed to degradation of soil capital through erosion, compaction, and localised salinisation [37, 38].

Many land parcels within the study landscape have come under the ownership and management of a single entity and are part of a 25-year plan to extensively reconfigure land uses for improved agricultural production and for improved environmental stewardship. Associated changes in the location and types of agricultural and restoration practices will directly and indirectly affect the supply of soil services [31], making this an ideal landscape for examining service changes under likely future scenarios. The 25-year plan is specific to 6,441 ha, a subset of the 30,200 ha study area, and includes proposed land uses on a continuum from less intensively managed systems ("ecological estate" with and without minor grazing and eucalypt plantation) to more intensively managed systems (mainly irrigated cropping, limited perennial pasture, and perennial horticulture; Figure 2).

Soil Sampling and Analysis

We used a 2 km grid as the starting point for the soil sampling design on the basis that grid sampling is often recommended when ordinary kriging is to be used for spatial distribution predictions [28]. Such a sampling scheme minimises the kriging variance, that is, the standard error of prediction [39, 40]. A 2 km sampling grid was proposed in this study because of its capacity to provide an adequate sampling intensity within cost and logistical constraints. When overlain on the study landscape, the 2 km grid yielded c. 60 strata (segments) that coincided with accessible land parcels covered by the abovementioned 25-year plan. One sampling point was chosen at the centre of each of these strata or from the nearest available location, leading to an irregular-grid sampling scheme consisting of 60 sampling points (Figure 2).

Each sampling point was located in the field, and an area of 10 by 10 m was defined for sampling. Within each sampling-point area, three randomly located subsamples were collected from each of three soil depths (0–10, 10–20, and 20–30 cm) using a soil auger of 8.25 cm internal diameter (AMS Inc., ID, USA). Subsamples from each depth were bulked to give a single composite sample per depth per sample-point area. Separate intact cores (custom-made 7.5 × 10 cm core) were also collected for bulk density determination.

The 180 soil samples (60 sampling points × 3 depths) were air-dried and passed through a 2 mm sieve in preparation for mid-infrared (MIR) spectral analysis. A subset of 60 soil samples were analysed for selected chemical and physical properties (see Table 1), and MIR spectroscopy was then used to estimate these properties in the remaining 120 samples. For MIR spectral scanning, a sub-sample (approximately 7 g) of each soil was finely ground to an approximate particle size of less than 0.1 mm diameter using a vibrating mixer ball mill (Retsch, Haan, Germany) set on a vibrational frequency of 20 s^{-1} for 3 min.

Table 1: Analysed soil chemical and physical properties of 60 sampling points (n=60) at each of three depths from the study area in northern Victoria, south-eastern Australia. Data were produced through conventional analysis methods as well as mid-infrared spectroscopy*

Soil properties	Min	Max	Mean	SD	Median
Organic carbon (%)a					
0–10 cm	0.3	4.5	1.3	0.8	1.2
10–20 cm	0.2	2.4	0.8	0.4	0.7
20–30 cm	0.1	1.6	0.6	0.3	0.5
pH_{CaCl_2} b					
0–10 cm	5.4	8.7	7.1	0.7	7.1
10–20 cm	5.6	9.3	7.5	0.7	7.6

20–30 cm	5.2	9.3	7.7	0.9	7.8
EC (dS/m)c					
0–10 cm	0.04	5.46	0.68	1.01	0.27
10–20 cm	0.05	9.80	1.09	1.62	0.49
20–30 cm	0.04	10.72	1.60	2.08	0.75
Clay (%)d					
0–10 cm	6	60	40	10	42
10–20 cm	11	61	43	10	46
20–30 cm	6	61	46	12	49
Sand (%)d					
0–10 cm	25	93	48	13	45
10–20 cm	17	84	43	14	38
20–30 cm	16	93	39	17	35
Bulk density (g/cm3)					
0–10 cm	1.0	1.6	1.3	0.1	1.3
10–20 cm	1.2	1.6	1.4	0.1	1.4
20–30 cm	1.2	1.6	1.4	0.1	1.4

[a]Dry combustion and Walkley-Black methods [41].

[b]pH in 0.01 M $CaCl_2$, 1 : 5 extraction ratio [42].

[c]EC in 1 : 5 water extraction ratio [43].

[d]Hydrometer method [44].

*Conventional analysis methods were applied to 61 reference samples (i.e., 61 out of the 180 study samples), and MIR spectroscopy was used to predict properties in the remaining samples.

Analysis of Soils Using MIR Spectroscopy and Partial Least-Squares (PLS) Regression

We used a Spectrum-One (PerkinElmer, Wellesley, MA, USA) Fourier transform MIR spectrometer to collect diffuse reflectance MIR spectra of the complete set of 180 soils following a methodology described elsewhere [45, 46]. Soil MIR spectroscopy provides the possibility of rapid, inexpensive, and simultaneous characterisation of various chemical properties in large numbers of samples [2, 45, 47]. Moreover, the repeatability and reproducibility of this technique among different laboratories have been found to be superior to the performance of conventional soil analyses [2]. prior to partial least-squares (PLS) regression, principal component analysis (PCA) was performed to detect any spectral outliers and to select a calibration subset of 60 soil samples for PLS model development. PCA revealed one sample as a spectral outlier, which along with the calibration subset was included in the reference soil property characterisation (Table 1).

The calibration models for predicting SOC, pH, electrical conductivity (EC), and percent sand, silt, and clay were developed between MIR spectra (predictor variables) and reference soil property values (response variables) based on the method introduced by Haaland and Thomas [48], which is popular in quantitative soil MIR analysis [47]. The softwares GRAMS/AI, GRAMS IQ, and IQ Predict, which are part of the GRAMS spectroscopy software suite (Thermo Fisher Scientific, Waltham, MA), were used for PCA and PLS-regression model building and predictions. The MIR-PLS prediction models from the calibration subset were applied to the remaining 119 samples (see Forouzangohar et al. [45, 46]), and these predicted soil properties were used in the remainder of the study.

APSIM for Predicting Indicators to Represent Priority Services

Two priority soil ecosystem services were identified for consideration, namely, carbon sequestration and water storage. These two services were nominated by the landholder as central to their land-use decision making (e.g., potential involvement in both carbon and water markets) and were identified as priority services for the benefit of humans and the environment by Aitkenhead et al. [14]. We chose to estimate current and future capacity to supply these services using two indicators, namely, SOC content and soil water storage. In this case study, the service "water storage" can include water added as irrigation. As such, this service represents capacity of the soil to retain a scarce resource (water), either from natural or artificial sources, in the landscape, and to reliably provide that resource for plant growth.

Key APSIM modules used in this study were SOILWAT [20, 49] for simulating soil water dynamics and SOILN [15, 19] for soil organic carbon. In addition, the MANAGER components were utilised to control land-use configurations (as below), and the plant growth modules (Agpasture, Soybean, Oat, Wheat, Lucerne, andEucalyptus grandis) were used to simulate the production systems corresponding to current and future land uses.

Two scenarios of indicator predictions were examined for the study landscape: one based on the current distribution of land uses and another based on a proposed future land-use configuration over the next 25 years (Figure 2). To isolate the effects of land-use change, climate change over the 25-year period was assumed to be negligible although some models predict lower rainfall and higher temperatures [50] in the study area, which could affect our indicator predictions. Data from a recent 25-year period (1985 to 2009) were used to represent climate in 25-years-time. In addition, to further avoid effects of interannual climate variability, daily weather data were defined for a "typical year" comprised of 12 "typical" months, each based on the average monthly rainfall and the distribution of rain days from two nearby weather stations (Swan Hill and Kerang, located, resp., to the north and south of the study area). Daily weather data for these weather stations were sourced from the SILO climate database (http://www.longpaddock.qld.gov.au/silo/) hosted by the Queensland Climate Change Centre of Excellence [51].

To represent the current land-use scenario (time = 0 years), APSIM was configured to simulate two land uses:

- dry cropping system, as a continuous winter cereal cropping simulated by the Wheat module, with removal of soil surface plant residues 30–45 days after harvest in preparation for opportunistic sowing in late autumn (leaving the soil surface bare for 4-5 months),
- pasture/grassland under intensive grazing, simulated by the Agpasture module, with regular fortnightly grazing to a remaining herbage of 500 kg/ha.

In addition, to represent the future land-use scenario (time = 25 years), APSIM was configured to simulate five new land uses (Figure 2):

- ecological estate (protected for restoration of perennial native vegetation, predominantly grasslands and shrublands), simulated by the Agpasture module, with no plant biomass removal,
- ecological estate with limited grazing, simulated by the Agpasture module, with two grazing events per year to a remaining herbage of 1000 kg/ha,

- irrigated cropping system, as summer-winter rotational cropping systems, simulated by the Soybean, Oats, and Wheat modules of crops, assuming no-tillage systems, and including no removal of soil surface plant residues after harvest,
- irrigated permanent lucerne stand (3 year ley), simulated by the Lucerne module, assuming regular harvest at flowering and removal of 95% of harvested plant material,
- eucalypt plantation, simulated by the Egrandis and Agpasture modules using the Canopy module to account for light and water competition between trees and intertree herbage (including no removal of plant biomass); note that APSIM currently lacks modules for tree species other than eucalypts, so, we treated limited areas of perennial horticulture as eucalypt plantations for the purposes of this study (i.e., perennial horticulture is not considered further).

Five soil types of the Mallee region in north-western Victoria, Australia, were chosen from the APSIM soil database to underpin APSIM modelling. The soils were chosen from those generic sites that had similar characteristics to sampled soils and were further parameterised for SOC, EC, pH, particle size distribution, and bulk density using our measures of the 60 reference soils.

Mapping Service Indicators

Ordinary kriging was used to estimate spatial distribution of service indicators over the study area under both current and future land-use scenarios. We used ArcGIS 10.0 (ESRI Inc., CA, USA) to produce distribution maps of indicators (SOC content and soil water storage) based on kriging between the 60 sample points. Here, the average distance between neighbouring sample points was 1300 m, so that a lag size of 1300 m was used in the semivariogram models (which depict the spatial autocorrelation of the sample points and define the weights of the kriging functions). With 10 lags of this size, semivariograms of up to 13,000 m were fitted for each service indicator under each scenario. The exponential functions were best fitted through semivariogram modelling, and nuggets were set to zero (i.e., it was assumed that there was no error for model-derived values at each sample point). Using the trend analysis tool, second-order polynomial trends were identified and fitted to remove directional trends in SOC content data. As required, log transformations were used to better approximate normal distributions of input data. Given a 2 km sampling grid, a range of 2,300 m was identified as appropriate when fitting the semivariogram models in all cases. In spatial interpolation of SOC content, each kriging average was obtained using a four-sector searching neighbourhood, searching for a maximum of 5 and a minimum of 2 points in each sector. In mapping soil water storage any points beyond the 2300 m range was not involved in predictions.

Mapping Hotspots of Service Change

Changes in the two service indicators between current and future scenarios were normalized to a 0-1 scale through linear scaling [52]. This allowed comparison of relative estimated changes in the two priority services, with a combined value of −2 indicating maximum negative change in both and of +2 indicating maximum positive change in both. These combined values were then used to produce a "hotspot" map (where a hotspot indicates positive change in both [30]) encompassing the 60 sampling points.

2.8. Assessing Uncertainties of Spatial Interpolations

The overall reliability of the indicator prediction surfaces (maps) over the entire study area was assessed using the mean prediction error and the standardized Root-Mean-Square Error (standardized RMSE). A mean prediction error close to zero (i.e., relative to the magnitude of the average prediction error) and a standardized RMSE close to one indicate relatively unbiased predictions of service supply/change. The reliability of the kriging model predictions for specific points over the study area was expressed as the ratio of performance deviation (RPD), which is the ratio of natural variation in the sample set (standard deviation, SD) to the size of the average prediction error (i.e., the higher the RPD, the more reliable the prediction [53]). An RPD value of <1.4 indicates poor performance, RPD > 1.4 indicates acceptable model performance for qualitative to semiquantitative analysis, and RPD > 2.0 indicates excellent performance [54].

RESULTS AND DISCUSSION

Soil Properties

There was significant variation in measured soil properties across the study area (Table 1). For example, SOC concentration to 10 cm depth ranged from 0.5 to 4.5% over distances of c. 10 km in the northern sections. Generally, SOC concentrations were greatest in northern and eastern sections (>2%) and least in central sections (<1%). The pH values ranged from slightly acidic (5.2 at 30 cm depth) to alkaline (9.3 at 30 cm depth), and EC ranged from 0.04 to 10.7 dS/m, reflecting patterns of localised surface salinization. Similarly, soil texture varied across the study area, with percent clay in the top 30 cm ranging from 6 to 61% and sand from 16 to 93% (Table 1).

Service Indicators under Alternative Land Uses

Simulations over a twenty-five-year period for a predominant clay soil in the study area predicted strong potential for increases in service indicators under the five new land uses. For example, mean SOC content of 30 t/ha under current land uses was predicted to increase under four of the new land uses, the highest being an estimated 38 t/ha under irrigated no-till cropping (Table 2). Increases in mean soil water storage were only predicted where new land uses involved irrigation (Table 2). However, increases in the ratio of aboveground biomass to soil water storage were predicted from 0.8 and 0.9 kg/m^3 under current land uses to 2.2 kg/m^{3}under new irrigated land uses and to 2.3, 3.9, and 84.5 kg/m^3 for nonirrigated new land uses of ecological estate with grazing, ecological estate without grazing, and eucalypt plantation, respectively (Table 2).

Table 2: Predicted service indicators under both current and new land uses for a predominant clay soil in the study landscape. Values were derived from 25-year simulations based on average properties of the clay soil to 30 cm depth

	Soil organic carbon (t/ha)	Soil water storage (m3/ha)	Aboveground biomass to soil water storage
Current land uses			
Dry cropping	30b	760	0.8
Intensive grazing	30b	680	0.9
Future land uses			
Ecological estate	34	710	3.9
Ecological estate with grazing	33	690	2.3
Eucalypt plantation	34	750	84.5
Irrigated no-till cropping	38	1200c	2.2
Irrigated permanent Lucerne	27	1170c	1.1

[a]Aboveground biomass data is derived from APSIM simulations.

[b]Initial soil organic carbon content in equilibrium (modelling assumption).

[c]Soil water topped up by irrigation.

Soil Organic Carbon Content: Current Status and Future Projections

On average, soil to 30 cm depth at the 60 sampling points under current management stored 35 t/ha soil organic carbon ("SOC"; range 13–102 t/ha; Table 3). The greatest current SOC stores (>55 t/ha) were at the northern end of the study landscape (sampling points W1, W2, W3, W4, and W7; see Figure 2 for point locations) where sampling points were adjacent to remnant native eucalypt woodlands along the Little Murray River. Isolated pockets of low relative carbon stores (<20 t/ha) were scattered throughout the study area and were associated with general cropping, mixed farming, and grazing (W8, W10, W12, W18, W36, R10).

Table 3: Estimates of service indicator at 60 sampling points (0–30 cm depth) and over the entire study area, under current and future (25-year) land-use scenarios. Overall changes in each indicator were based on differences between supply on an average day under the current scenario and on an average day in 25-year-time under the future scenario

Indicator	Sampling points				Entire study areaa
	Min	Max	Mean	SD	
Soil organic carbon (t/ha)					
Current	13	102	35	17	165,119 t
Future	18	92	39	14	183,057 t
Change	−11	11	3	4	17,938 t
Soil water storage (m3/ha)					
Current	179	843	587	156	2,745,342 m3
Future	189	1220	719	251	3,362,834 m3
Change	0	463	132	146	617,492 m3

[a]Study area of 6,441 ha.

Predictions of soil carbon change under the future land-use scenario indicated a mean increase over the 25-year period of 3 t/ha to 30 cm depth at the 60 sampling points (Table 3). Over the 6,441 ha covered by the 25-year plan, this translated to an increase of 17,938 t SOC (i.e., change from 165,119 to 183,057 t; Table 3). Simulations indicated a range of potential changes in SOC of up to ±11 t/ha (Figure 3(a), Table 3).

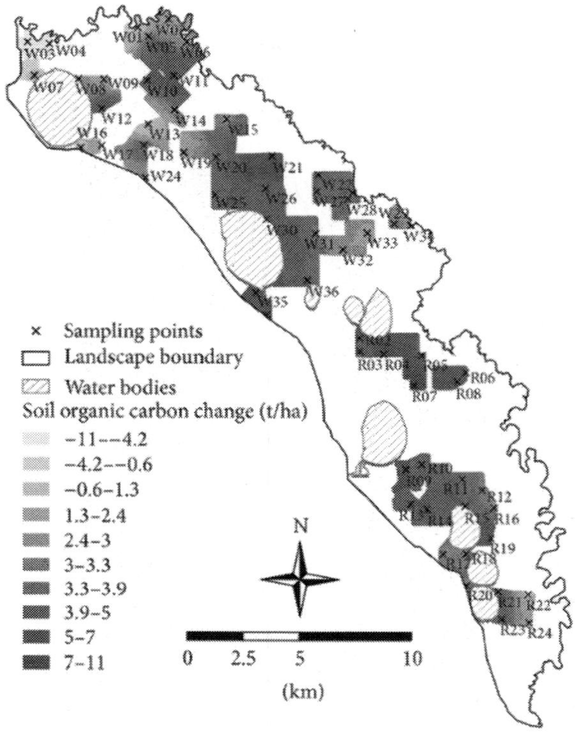

Figure 3: Predicted changes in service indicators, (a) soil organic carbon content and (b) soil water storage, to 30 cm depth in the study area associated with changes from current (time = 0) to future (time = 25 years) land-use scenarios. Each prediction surface represents differences in the supply of services between an average day under the current scenario and an average day under the future scenario. (a) Mean prediction error: 0.04; standardized RMSE: 1.06; average prediction error: 2.4; RPD: 1.5; (b) mean prediction error: −4.8; standardized RMSE: 1.09; average prediction error: 127; RPD: 1.2.

Spatial predictions of SOC changes were acceptably reliable (RPD > 1.4; Figure 3(a)) and highlighted potential risks and opportunities in soil carbon management [25, 30]. For example, decreases of up to 11 tC/ha were predicted near the northern end (Figure 3(a)), corresponding with the greatest current SOC stores (>55 t/ha). Reasons for the high relative current SOC here were unclear, although they could relate to the close proximity to remnant perennial vegetation and/or to time since clearing (unknown). Nonetheless, this study's predictions indicate that the planned change in land use from dry cropping to zero-tillage irrigated cropping systems will not prevent SOC decreases, and that alternative land uses should be examined to conserve this carbon content if sustaining SOC content is a priority. Elsewhere, predicted decreases of c. 4 tC/ha near sampling point W13 in the northern section (Figure 3(a)) were associated with a planned land-use change from regular grazing to an irrigated lucerne system, involving regular removal of plant material from the soil surface. Given the already low carbon stores at this sampling point (28.4 t/ha), alternative land uses that retain rather than remove plant material might be considered.

Landscape locations that show high potential increases in SOC might also be considered for targeted management. For example, locations of maximum predicted increases (e.g., R7, W21, and W35; Figure 3(a)) were associated with changes from intensive agriculture to a zero-tillage irrigated cropping system with no plant residue removal. These data thus support strong recommendations for adoption of no-till farming by Lal et al. [55] as an option for enhancing SOC storage in cropland soils of relatively low current SOC content (<30 tC/ha). Another example of an opportunity to increase SOC stores was indicated at sampling point W12. Here, the soil was a loamy sand of very low SOC content (13.2 t/ha), and a land-use change from mixed farming with intensive grazing to ecological estate with limited grazing led to predicted increases of 5.7 tC/ha (43%) over 25 years (Figure 3(a)).

Overall, changes from intensive agriculture to ecological estate, ecological estate with limited grazing, and eucalypt plantations were associated with moderate increases in SOC of 1.3 to 5 tC/ha over the 25-year period (Figure 3(a)). This is consistent with previous findings of measured SOC increases with conversion from traditional croplands to perennial grasslands (e.g., [56–58]). Kätterer et al. [56] reported an average of 0.4 t/ha/yr increase in SOC content over 30 years when a cropland was converted back to grassland in a Swedish farm, although their study involved a wetter (542 mm/yr) and cooler (mean annual temperature of 5.1°C) climate and double the starting SOC content. In this study's landscape, limited rainfall and very hot and dry summers are likely to be limiting factors for grassland growth, as simulated by the APSIM-Agpasture module. Nonetheless, in terms of percent gain in SOC through conversion of degraded lands to natural grasslands, our results show agreement with previously reported limits of 15–20% increases [56, 57, 59].

Soil Water Storage: Current Status and Future Projections

Current water storage of soils reflected the predominance of clay-textured soils in the 6,441 ha covered by the 25-year plan. Total water stored to 30 cm soil depth on an average day under the current scenario was estimated at 2,745,342 m^3 (Table 3). On average, soils stored 587 m^3/ha water without irrigation. This value ranged from 179 m^3/ha for a loamy sand soil under regularly-grazed pasture (sample point W12) to 843 m^3/ha in a clay soil under dryland cereal cropping (sample point R7).

Land-use changes in the 25-year plan were predicted to increase water storage by 22% (i.e., to a mean water storage of 719 m^3/ha and a total storage capacity of 3,362,834 m^3; Table 3). Similarly, modelled changes in average daily soil water storage over the entire study landscape ranged from 0 to 463 m^3/ha but were mostly >110 m^3/ha (Figure 3(b)). Modelled soil water storage values under current and future land-use scenarios were consistent with APSIM-generated published works [21, 22]. For instance, volumetric soil water in the top 30 cm of selected soil profiles was predicted and observed to be between 0.1 and 0.4 water volume/total volume (c. 300 to 1200 m^3/ha) in the nearby Murray-Darling Basin [22].

Predicted increases in soil water storage were associated with changes to irrigated no-till systems, and, to a lesser degree, increases in perenniality. Irrigated no-till cropping systems add water from outside the system and can increase water retention through a mulching effect associated with retained plant residues. Mean predicted increases in mean water storage in soils under this land-use change were 54%, from 669 m^3/ha to 1,029 m^3/ha (i.e., deepest red areas for points R5, R7-R9, R14, R17, R19, W3, W4, W7, W9, W11, W21, and W35 in Figure 3(b)). Increases in soil water storage over the 25-year period were also indicated for land uses that increased perenniality, namely, eucalypt plantation and ecological estate (with or without grazing). In these land uses, increases of 8–15% in average daily water storage were apparently due to processes affecting water infiltration and retention and were perhaps tempered by increased losses through evapotranspiration associated with the establishment of woody species [60]. This is also consistent with studies that have indicated that the reverse land-use change (i.e., from woody native vegetation to cropping) was associated with decreased soil water-holding capacity [61].

Relationships among Soil Ecosystem Services

The indicator change maps highlight locations both of potential complementarities (synergies) among priority services, as indicated by mutual increases with changed management, and of potential trade-offs among services, as indicated by marked increases in one service but decreases in another. Overall, we found mostly positive (or "synergic"; [62]) relationships among our priority soil services (Figure 4). This suggests that they increased at the same time, due either to simultaneous responses to the same land-use change or to positive interactions [62]. In this context, hotspots are located at points where the supply of multiple soil services can be mutually and markedly improved [30].

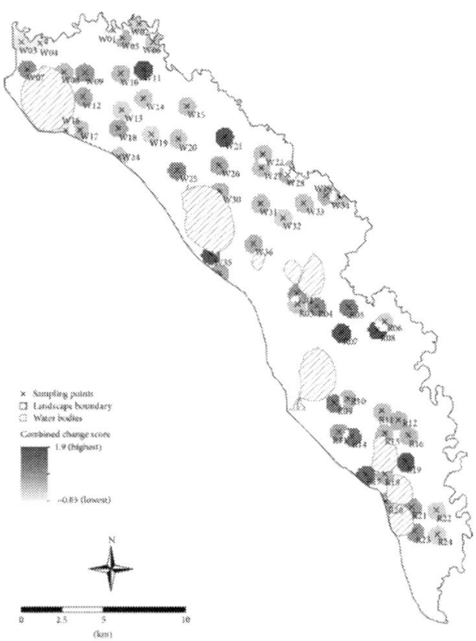

Figure 4: "Hotspot" map indicating combined relative change in the two service indicators (soil organic carbon content and soil water storage) from current (time = 0) to future (time = 25 years) land-use scenarios at each of

60 sampling points. Potential combined scores range from −2 (maximum negative change in both indicators) to +2 (maximum positive change in both indicators).

This study found several potential hotspots as indicated by high combined service change scores (>1 to 2; Figure 4). For example, land-use change to irrigated no-till cropping led to substantial estimated increases in both soil water storage and SOC content at sampling points R7, R8, R9, R14, R17, R19, W11, W21, and W35 (Figure 4). However, the study also highlighted at least six instances of possible trade-offs between services as indicated by very low combined change scores (<0.2). Four of these cases were located at sampling points W1, W3, W4, and W13, where future predictions flagged likely reductions in SOC stocks. However, we doubt this is due to any causal relationships among the services, but, as discussed in Section 3.3, this appears to be due to the relatively high initial SOC contents at these points (except W13), which were not sustained by the new land uses. Similarly, the low combined change score at W13 was mainly due to predicted decreases in SOC associated with regular removal of plant material under the planned irrigated lucerne system. Elsewhere, the low combined change score at W16 could be attributed to negligible change in land-use practices and thus in service provision under unchanged climatic conditions.

Approach Limitations

We acknowledge limitations in our approach. Our spatial interpolation method of choice, kriging from point data, carries some limitations. Prediction statistics were acceptable for SOC (mean prediction error 0.04, standardised RMSE 1.06, and RPD 1.5) but were less convincing for soil water storage (mean prediction error −4.8, standardised RMSE 1.09, and RPD 1.2). In particular, the low RPD for soil water storage indicated nonreliable estimations for at least some locations perhaps due to greater spatial variation in soil water distributions than SOC.

In addition, the APSIM model could not equally represent our diversity of land-use changes. For example, APSIM plant modules have not been specifically developed for diverse native plant systems (i.e., our "ecological estate") and for perennial horticulture systems. Moreover, it was not feasible to fully parameterise the APSIM soil components for the 60 sampling points in this study. Instead, we relied on the APSIM soil database to provide default values for some of our modelling components. This leads to uncertainties, the quantification of which is beyond the scope of this study. Future work will examine the implications of various assumptions, used not only in the APSIM modelling but also in the entire approach, through detailed sensitivity analyses.

CONCLUSIONS

We demonstrated a broadly applicable approach for quantifying and mapping service indicators (representing priority soil ecosystem services) in an agricultural landscape in south-eastern Australia. A feasible sampling intensity was combined with soil MIR spectral analysis, biophysical modeling, and spatial interpolations to provide estimates of two soil indicators under both current and future land-use scenarios. We showed that under the future land-use plan, the supply of the two priority soil services (based on indicators) would likely increase. The modelled maps provide a basis for supporting decisions about alternative land uses by indicating "hotspot" locations where there are mutual increases in the supply of soil ecosystem services or locations where there is a risk to the supply of some services. Such knowledge informs the management of multiple services at a range of scales from paddock to landscape and region, both in response to markets for particular services and to requirements or incentives for providing multiple benefits (see, e.g., the proposed "cobenefits index" in Australia's recent "Carbon Credits Bill"; Commonwealth of Australia, 2011).

The proposed approach could be useful in designing strategies for supporting sustainable soil management in this landscape. Coordinated management of multiple land parcels within a landscape, rather than isolated management of single parcels, offers potential to maintain supply of priority services at the landscape scale by avoiding risks and harnessing opportunities through matching soils with particular land-use practices. Introducing new land uses in any given landscape not only offers scope for increasing landscape complexity [63, 64] but also offers more flexibility in decisions relating to landscape-level delivery of multiple services. This type of flexibility and responsiveness will be needed as additional factors like climate change further confound the challenge to managing soils sustainably in changing agricultural landscapes.

ACKNOWLEDGMENTS

This study was funded by the (then) Land Health Unit, Natural Resources Division of the (now) Victorian Department of Environment and Primary Industries. The authors would also like to thank Kilter Pty Ltd., in particular Malory Weston and David Heislers, for providing them with scenario information and access to properties and for help with soil sampling; APSIM development team for providing them with APSIM software and technical advice; Rai Kookana, Mike McLaughlin, and Les Janik from CSIRO Land and Water, Adelaide for access to the soil MIR spectroscopy facility; and Narges Milani, Saravanan Jangammanaidu Krishnaraj, Tarek Murshed, Jane Lean, Mathew Lee, Najib Ahmady, and Lyn Wilks from the Department of Forest and Ecosystem Science, the University of Melbourne, for their help and technical assistance in the collection and analysis of soil samples.

REFERENCES

1. E. Dominati, M. Patterson, and A. Mackay, "A framework for classifying and quantifying the natural capital and ecosystem services of soils," Ecological Economics, vol. 69, no. 9, pp. 1858–1868, 2010.
2. C. Palm, P. Sanchez, S. Ahamed, and A. Awiti, "Soils: a contemporary perspective," Annual Review of Environment and Resources, vol. 32, pp. 99–129, 2007.
3. D. A. Robinson, N. Hockley, E. Dominati et al., "Natural capital, ecosystem services, and soil change: why soil science must embrace an ecosystems approach," Vadose Zone Journal, vol. 11, no. 1, 2012.
4. D. A. Robinson, I. Lebron, and H. Vereecken, "On the definition of the natural capital of soils: a framework for description, evaluation, and monitoring," Soil Science Society of America Journal, vol. 73, no. 6, pp. 1904–1911, 2009.

5. S. Frank, C. Fürst, L. Koschke, and F. Makeschin, "A contribution towards a transfer of the ecosystem service concept to landscape planning using landscape metrics," Ecological Indicators, vol. 21, pp. 30–38, 2012.
6. M. Rutgers, H. J. van Wijnen, A. J. Schouten et al., "A method to assess ecosystem services developed from soil attributes with stakeholders and data of four arable farms," Science of the Total Environment, vol. 415, pp. 39–48, 2012.
7. H. J. van Wijnen, M. Rutgers, A. J. Schouten, C. Mulder, D. de Zwart, and A. M. Breure, "How to calculate the spatial distribution of ecosystem services—natural attenuation as example from The Netherlands," Science of the Total Environment, vol. 415, pp. 49–55, 2012
8. R. Haines-Young, M. Potschin, and F. Kienast, "Indicators of ecosystem service potential at European scales: mapping marginal changes and trade-offs," Ecological Indicators, vol. 21, pp. 39–53, 2012.
9. L. T. Bennett, P. M. Mele, S. Annett, and S. Kasel, "Examining links between soil management, soil health, and public benefits in agricultural landscapes: an Australian perspective," Agriculture, Ecosystems & Environment, vol. 139, no. 1-2, pp. 1–12, 2010.
10. R. W. Simonson, "Outline of a generalized theory of soil formation," Soil Science Society of America Proceedings, vol. 23, no. 2, pp. 152–156, 1959.
11. B. Burkhard, F. Kroll, S. Nedkov, and F. Müller, "Mapping ecosystem service supply, demand and budgets," Ecological Indicators, vol. 21, pp. 17–29, 2012.
12. B. Minasny and A. E. Hartemink, "Predicting soil properties in the tropics," Earth-Science Reviews, vol. 106, no. 1-2, pp. 52–62, 2011
13. M. U. F. Kirschbaum, J. O. Carter, P. R. Grace et al., "Brief description of several models for simulating net ecosystem exchange in Australia," in Net Ecosystem Exchange, M. U. F. Kirschbaum and R. Mueller, Eds., CRC for Greenhouse Accounting, Canberra, Australia, 2001.
14. M. J. Aitkenhead, F. Albanito, M. B. Jones, and H. I. J. Black, "Development and testing of a process-based model (MOSES)

for simulating soil processes, functions and ecosystem services," Ecological Modelling, vol. 222, no. 20–22, pp. 3795–3810, 2011.

15. B. A. Keating, P. S. Carberry, G. L. Hammer et al., "An overview of APSIM, a model designed for farming systems simulation," European Journal of Agronomy, vol. 18, no. 3-4, pp. 267–288, 2003.
16. R. L. McCown, G. L. Hammer, J. N. G. Hargreaves, D. P. Holzworth, and D. M. Freebairn, "APSIM: a novel software system for model development, model testing and simulation in agricultural systems research," Agricultural Systems, vol. 50, no. 3, pp. 255–271, 1996.
17. R. J. Farquharson, G. D. Schwenke, and J. D. Mullen, "Should we manage soil organic carbon in Vertosols in the northern grains region of Australia?" Australian Journal of Experimental Agriculture, vol. 43, no. 3, pp. 261–270, 2003.
18. Z. Luo, E. Wang, O. J. Sun, C. J. Smith, and M. E. Probert, "Modeling long-term soil carbon dynamics and sequestration potential in semi-arid agro-ecosystems," Agricultural and Forest Meteorology, vol. 151, no. 12, pp. 1529–1544, 2011.
19. K. Ranatunga, M. J. Hill, M. E. Probert, and R. C. Dalal, "Comparative application of APSIM, RothC and Century to predict soil carbon dynamics," in Proceedings of the International Congress on Modelling and Simulation (MODSIM ‹01), F. Ghassemi, P. Whetton, R. Little, and M. Littleboy, Eds., pp. 733–738, Modelling and Simulation Society of Australia and New Zealand, Canberra, Australia, 2001.
20. B. A. Keating, D. Gaydon, N. I. Huth et al., "Use of modelling to explore the water balance of dryland farming systems in the Murray-Darling Basin, Australia," European Journal of Agronomy, vol. 18, no. 1-2, pp. 159–169, 2002.
21. M. E. Probert, J. P. Dimes, B. A. Keating, R. C. Dalal, and W. M. Strong, "APSIM›s water and nitrogen modules and simulation of the dynamics of water and nitrogen in fallow systems," Agricultural Systems, vol. 56, no. 1, pp. 1–28, 1998.
22. J. Wang, E. Wang, and D. L. Liu, "Modelling the impacts of climate change on wheat yield and field water balance over the Murray-Darling Basin in Australia," Theoretical and Applied Climatology, vol. 104, no. 3-4, pp. 285–300, 2011.

23. R. O. Gilbert and J. C. Simpson, "Kriging for estimating spatial pattern of contaminants: potential and problems," Environmental Monitoring and Assessment, vol. 5, no. 2, pp. 113–135, 1985.
24. A. B. McBratney and R. Webster, "How many observations are needed for regional estimation of soil properties?" Soil Science, vol. 135, no. 3, pp. 177–183, 1983.
25. P. Longley, Geographic Information Systems & Science, John Wiley & Sons, Hoboken, NJ, USA, 2011.
26. Z. L. Frogbrook, J. Bell, R. I. Bradley et al., "Quantifying terrestrial carbon stocks: examining the spatial variation in two upland areas in the UK and a comparison to mapped estimates of soil carbon," Soil Use and Management, vol. 25, no. 3, pp. 320–332, 2009.
27. S. Kumar and R. Lal, "Mapping the organic carbon stocks of surface soils using local spatial interpolator," Journal of Environmental Monitoring, vol. 13, no. 11, pp. 3128–3135, 2011.
28. J. W. van Groenigen, W. Siderius, and A. Stein, "Constrained optimisation of soil sampling for minimisation of the kriging variance," Geoderma, vol. 87, no. 3-4, pp. 239–259, 1999.
29. C. S. Zhang, Y. Tang, X. L. Xu, and G. Kiely, "Towards spatial geochemical modelling: use of geographically weighted regression for mapping soil organic carbon contents in Ireland," Applied Geochemistry, vol. 26, no. 7, pp. 1239–1248, 2011.
30. N. D. Crossman and B. A. Bryan, "Identifying cost-effective hotspots for restoring natural capital and enhancing landscape multifunctionality," Ecological Economics, vol. 68, no. 3, pp. 654–668, 2009.
31. P. M. Haygarth and K. Ritz, "The future of soils and land use in the UK: soil systems for the provision of land-based ecosystem services," Land Use Policy, vol. 26, no. 1, pp. S187–S197, 2009.
32. M. Schröter, R. P. Remme, and L. Hein, "How and where to map supply and demand of ecosystem services for policy-relevant outcomes?" Ecological Indicators, vol. 23, pp. 220–221, 2012.
33. R. F. Isbell, The Australian Soil Classification, CSIRO Australia, Victoria, Australia, 2002.
34. DPI, "Geomorphology of Victoria (GMU250/GMU250)," Department of Primary Industries, Bendigo, Victoria, Australia, 2010.

35. Ramsar Convention Secretariat, The Ramsar Convention Manual: A Guide To the Convention on Wetlands (Ramsar, Iran, 1971), Ramsar Convention Secretariat, Gland, Switzerland, 2011.
36. DPI, "Victorian land use information system dataset," Department of Primary Industries, Bendigo, Victoria, Australia, 2009.
37. J. Connor, "The economics of time delayed salinity impact management in the River Murray," Water Resources Research, vol. 44, no. 3, 2008.
38. N. D. Crossman, J. D. Connor, B. A. Bryan, D. M. Summers, and J. Ginnivan, "Reconfiguring an irrigation landscape to improve provision of ecosystem services," Ecological Economics, vol. 69, no. 5, pp. 1031–1042, 2010.
39. G. Christakos and R. A. Olea, "Sampling design for spatially distributed hydrogeologic and environmental processes," Advances in Water Resources, vol. 15, no. 4, pp. 219–237, 1992.
40. A. W. Warrick and D. E. Myers, "Optimization of sampling locations for variogram calculations," Water Resources Research, vol. 23, no. 3, pp. 496–500, 1987.
41. D. W. Nelson and L. E. Sommers, "Total carbon, organic carbon, and organic Matter," in Methods of Soil Analysis—part 3: Chemical Methods, D. L. Sparks, Ed., pp. 961–1010, Soil Science Society of America, Madison, Wis, USA, 1996.
42. G. W. Thomas, "Soil pH and soil acidity," in Methods of Soil Analysis—part 3: Chemical Methods, D. L. Sparks, Ed., pp. 475–490, Soil Science Society of America, Madison, Wis, USA, 1996.
43. J. D. Rhoades, "Salinity: electrical conductivity and total dissolved solids," in Methods of Soil Analysis—part 3: Chemical Methods, D. L. Sparks, Ed., pp. 417–435, Soil Science Society of America, Madison, Wis, USA, 1996.
44. G. W. Gee and J. W. Bauder, "Particle-size analysis," in Methods of Soil Analysis—part 1: Physical and Mineralogical Methods, D. L. Sparks, Ed., pp. 383–411, Soil Science Society of America, Madison, Wis, USA, 1996.
45. M. Forouzangohar, D. Cozzolino, R. S. Kookana, R. J. Smernik, S. T. Forrester, and D. J. Chittleborough, "Direct comparison between visible near- and mid-infrared spectroscopy for describing diuron sorption in soils," Environmental Science & Technology, vol. 43, no. 11, pp. 4049–4055, 2009.

46. M. Forouzangohar, R. S. Kookana, S. T. Forrester, R. J. Smernik, and D. J. Chittleborough, "Midinfrared spectroscopy and chemometrics to predict diuron sorption coefficients in soils," Environmental Science & Technology, vol. 42, no. 9, pp. 3283–3288, 2008.
47. L. J. Janik, R. H. Merry, and J. O. Skjemstad, "Can mid infrared diffuse reflectance analysis replace soil extractions?" Australian Journal of Experimental Agriculture, vol. 38, no. 7, pp. 681–696, 1998.
48. D. M. Haaland and E. V. Thomas, "Partial least-squares methods for spectral analyses. 1. Relation to other quantitative calibration methods and the extraction of qualitative information," Analytical Chemistry, vol. 60, no. 11, pp. 1193–1202, 1988.
49. K. Ranatunga, E. R. Nation, and D. G. Barratt, "Review of soil water models and their applications in Australia," Environmental Modelling & Software, vol. 23, no. 9, pp. 1182–1206, 2008
50. B. Timbal and D. A. Jones, "Future projections of winter rainfall in southeast Australia using a statistical downscaling technique," Climatic Change, vol. 86, no. 1-2, pp. 165–187, 2008
51. S. J. Jeffrey, J. O. Carter, K. B. Moodie, and A. R. Beswick, "Using spatial interpolation to construct a comprehensive archive of Australian climate data," Environmental Modelling & Software, vol. 16, no. 4, pp. 309–330, 2001
52. K. Swingler, Applying Neural Networks: A Practical Guide, Academic Press, San Francisco, Calif, USA, 1996.
53. T. Fearn, "Assessing calibrations: SEP, RPD, RER, and R^2," NIR News, vol. 13, no. 6, pp. 12–14, 2002
54. C.-W. Chang, D. A. Laird, M. J. Mausbach, and C. R. Hurburgh, "Near-infrared reflectance spectroscopy—principal components regression analyses of soil properties," Soil Science Society of America Journal, vol. 65, no. 2, pp. 480–490, 2001.
55. R. Lal, M. Griffin, J. Apt, L. Lave, and M. G. Morgan, "Managing soil carbon," Science, vol. 304, no. 5669, p. 393, 2004.
56. T. Kätterer, L. Andersson, O. Andrén, and J. Persson, "Long-term impact of chronosequential land use change on soil carbon stocks on a Swedish farm," Nutrient Cycling in Agroecosystems, vol. 81, no. 2, pp. 145–155, 2008.

57. M. P. McHenry, "Farm soil carbon monitoring developments and land use change: unearthing relationships between paddock carbon stocks, monitoring technology and new market options in Western Australia," Mitigation and Adaptation Strategies for Global Change, vol. 14, no. 6, pp. 497–512, 2009
58. L. P. Qiu, X. R. Wei, X. C. Zhang et al., "Soil organic carbon losses due to land use change in a semiarid grassland," Plant and Soil, vol. 355, no. 1-2, pp. 299–309, 2012.
59. L. B. Guo and R. M. Gifford, "Soil carbon stocks and land use change: a meta analysis," Global Change Biology, vol. 8, no. 4, pp. 345–360, 2002.
60. M. D. Nosetto, E. G. Jobbágy, and J. M. Paruelo, "Land-use change and water losses: the case of grassland afforestation across a soil textural gradient in central Argentina," Global Change Biology, vol. 11, no. 7, pp. 1101–1117, 2005.
61. G. Mahe, J.-E. Paturel, E. Servat, D. Conway, and A. Dezetter, "The impact of land use change on soil water holding capacity and river flow modelling in the Nakambe River, Burkina-Faso," Journal of Hydrology, vol. 300, no. 1–4, pp. 33–43, 2005.
62. E. M. Bennett, G. D. Peterson, and L. J. Gordon, "Understanding relationships among multiple ecosystem services," Ecology Letters, vol. 12, no. 12, pp. 1394–1404, 2009.
63. T. Tscharntke, J. M. Tylianakis, T. A. Rand et al., "Landscape moderation of biodiversity patterns and processes—eight hypotheses," Biological Reviews, vol. 87, no. 3, pp. 661–685, 2012.
64. J. Vandermeer and I. Perfecto, "The agricultural matrix and a future paradigm for conservation,"Conservation Biology, vol. 21, no. 1, pp. 274–277, 2007.

Chapter 6

Assessing Drivers of Plantation Forest Productivity on Eroded and Non-Eroded Soils in Hilly Land, Eastern North Island, New Zealand

Marie J Heaphy[1], David J Lowe[2], David J Palmer[3], Haydon S Jones[4], Gerty JHP Gielen[1], Graeme R Oliver[1], and Stephen H Pearce[1]

[1]Scion, Private Bag 3020, Rotorua, New Zealand

[2]School of Science, University of Waikato, Private Bag 3105, Hamilton 3240, New Zealand

[3]Landcare Research, Private Bag 3127, Hamilton 3240, New Zealand

[4]Waikato Regional Council, Private Bag 3038, Hamilton 3240, New Zealand

ABSTRACT

Background

The effect of soil erosion on New Zealand production forestry is not well known and there has been no research prior to our study into the relationship between soil nutrient status and planted forests growing in eroded soils in steeplands.

Methods

The impact of soil erosion by mass movement on forest productivity was investigated in a paired plot trial in a planted forest in a mainly hilly to steepland catchment (Pakuratahi) near Napier, eastern North Island, New Zealand. Tree growth and form were measured and soil properties analysed to compare productivity and productivity drivers in adjacent non-eroded and eroded plots.

Results

Regression analysis showed that the decreased soil total nitrogen, total carbon, total phosphorus, and soil organic matter content in eroded plots had a negative impact on tree volume, resulting in a 10% decrease in measured tree volume. Based on an assessment of log quality, trees in the eroded plots were forecast to produce 16% less volume from high-quality pruned logs (with associated reduction in revenue of around $4000 per hectare), than trees in non-eroded plots. The total recoverable volume (TRV), estimated (for a 25-year rotation) from the measured *Pinus radiata* D. Don trees growing on the eroded sites, was valued at $68,500, about 9% less than the estimated TRV from trees measured on non-eroded plots ($76,000). Tree form and mean tree height in eroded and non-eroded plots were not significantly different.

Conclusions

Soil erosion impacts production in planted forests. Afforestation of erodible land provides a valuable ecosystem service through land and soil stabilisation but this service is currently not reflected in the market prices for timber in New Zealand. Maintaining the productive capacity of erodible soils through practices such as fertilisation or continuous-cover forestry can add further costs to production forestry. To ensure that sustainable forest practices are carried out to protect the productivity of soils, financial incentives may be justified.

BACKGROUND

Erosion by mass movement is a common form of soil loss in steepland and hill country in New Zealand (Page et al. [1994]). The east coast of the North Island is predominantly a mudstone/sandstone terrain capped in places with tephra deposits. This area became particularly susceptible to accelerated erosion following settlement by European farmers in the late 19thCentury when much of the indigenous forest was removed and replaced with pasture. Subsequently, landsliding and other forms of soil erosion increased when subjected to intense rainfall events (Crozier [1986]; Glade [1998]; Page et al. [1994]). Mass movement in the form of soil slips comprises shallow, rapid slides and flows involving soil and regolith (Lynn et al. [2009]) leaving scars in source areas and debris tails (run-out deposits) in the landscape. The failure surface is typically planar and parallel to the ground surface and usually <1 m deep (soil slips>1 m deep are uncommon) (Lynn et al. [2009]). Soil in the debris tail may be deposited within the same land unit or transported off-site downstream. The eroded scar surface is slow to re-vegetate (often ≥10 years), and the rate of recovery is influenced by factors such as lithology, weathering rate (governing release of nutrients), fertility, water holding capacity, and climatic conditions (Blaschke et al. [2000]).

The effect of soil erosion on New Zealand production forestry is not well known and there has been no research prior to this study into the relationship between decreased soil nutrients and planted forests growing in eroded soils. Research on pasture reduction following erosion, however, has been covered by a number of trials. In eroded

Wairarapa hill country in the southern North Island of New Zealand, maximum production recovery occurred within 20 years but only to a level of 80% of the pre-eroded level (Rosser and Ross [2011]; Lambert et al. [1984], [1993]). Declines in productivity on hillslope pasture were measured at 2% per decade after forest removal (Trustrum et al. [1983]). Research into the effects of erosion on productivity in the United States has mainly focused on temperate, arable croplands. Of the few national and international studies researching the effects of erosion on forest productivity, all found a decrease in tree volume. In the east coast of New Zealand, earlier research[a] examined the forest productivity of first-rotation *Pinus radiata* D. Don (radiata pine) in 20 paired plots on known eroded and adjacent control sites. A 10% reduction in mean tree diameter at breast height was found in the slip-eroded areas in addition to 26% less total stem volume per hectare and a lower yield of more valuable pruned logs. In the Queen Charlotte Islands of British Columbia, 49 landslides ranging in age from 1 to 155 years were surveyed and it was found that the trees on landslides averaging an age of 85 years produced around one half the wood volumes of normal second-growth stands of the same age (Smith et al.[1986]). Landslides occurring in natural forests have been shown to have adverse effects on tree recovery, with reduced rates of biomass accumulation and a species composition different from that of the surrounding forest in Hawaii (Francescato et al. [2001]) and in New Hampshire, USA (Restrepo et al. [2003]).

[a]Dean, M, & Heron, C. (1998). The effect of historical soil erosion on tree quality and volume: Report No. 58, Forest and Farm Plantation Management Cooperative. Unpublished.

The values of avoided erosion through afforestation and the other ecosystem services that forestry provides have been quantified (Barry et al. [2014]) but these services are currently not reflected by market prices in New Zealand. Barry et al. ([2014]) concluded that where benefits to society outweigh costs to the forest owner, positive incentives should be legislated to promote afforestation of marginal land. In the current project, forest growth on erosion-prone soils was compared with tree growth in non-eroded soils in a plantation forest in mainly hilly land and steepland near Napier. The objectives were (1) to determine and compare soil nutrient status in eroded and non-eroded soils in a planted forest catchment (Pakuratahi) and to relate the findings to forest productivity, and (2) to estimate the decrease in timber values associated with the loss of soil by erosion in the Pakuratahi forest.

METHODS

This research was carried out in a northern sub-catchment of the Pakuratahi forest (Figure 1), which is managed by PanPac Forest Products Ltd. The Pakuratahi catchment is situated 21 km northwest of Napier, coastal Hawke's Bay, at latitude 39° 20′ 57″ and longitude 176° 52′ 30″ (Figure 1).

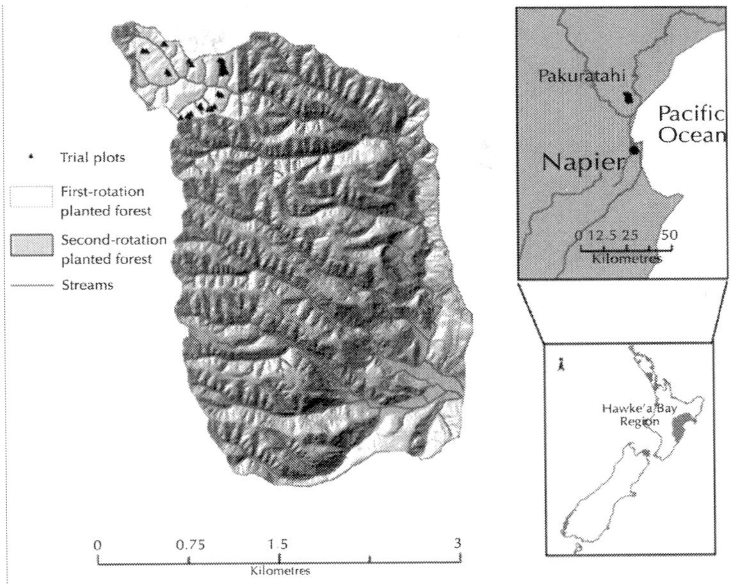

Figure 1: Location of study site,the pale-highlighted area in the northwestern section of Pakuratahi plantation forest in the Hawke's Bay region,New Zealand.

The Hawke's Bay region experiences a variable climate with summer droughts and high intensity rainfall events (Thompson [1987]). Mean annual rainfall ranges from 440 mm to 1300 mm with maximum daily rainfall ranging from 28 mm to 300 mm (NIWA [2010]). Mean annual rainfall and temperature at Tangoio, 5 km north of the Pakuratahi catchment, are 1501 mm and 13.3°C, respectively (Eyles and Fahey [2006]). Pakuratahi catchment comprises 774 ha of mainly hilly land and steepland (more than 60% of the catchment contains slopes of 20° or greater) with flat to moderately rounded hills or ridge tops, and narrow terraces in valley bottoms in the southern section of the catchment. Elevation ranges from 18 m to 355 m above sea level.

The main geological units underlying the Pakuratahi catchment are the Kaiwaka Formation, a Pleistocene sedimentary unit comprising calcareous sandstone, limestone, and conglomerate aged c. 1.7 million years and up to 100 m thick, and the underlying Petane Formation, an early Pleistocene succession of repetitive alternating sequences of limestone, sandstone, mudstone, and conglomerate dating back to c. 2.2 million years (Haywick et al. [1991]; Lee et al. [2011]). Tephra-derived soils, which mainly occur on the upper ridges and slopes, have developed *in situ*primarily in the Waimihia Tephra (deposited 3401 ± 108 calendar yr BP) and the Taupo Tephra (deposited 232 ± 10 AD) (Lowe et al. [2013]) over sandstone of the Kaiwaka and Petane formations (Eyles and Fahey [2006]). The main soil orders in the Pakuratahi catchment are Pallic Soils and Pumice Soils (New Zealand Soil Classification: Hewitt [2010]). Recent Soils also occur in the south of the catchment around the Pakuratahi River (Newsome et al. [2008]).

Site Selection

Data from a digitised surface of erosion slips photographed after storm events between 1943 and 1994 (Fransen and Brownlie [1995]) were used to select erosion scars for sampling. Erosion scars greater than 0.02 ha in area and covered by a 21-year-old first-rotation *Pinus radiata* plantation were selected and identified as suitable for field examination. Fifteen plots were established within the erosion scars and paired with 15 plots located on adjacent non-eroded (stable) land of similar aspect and slope (Figure 2).

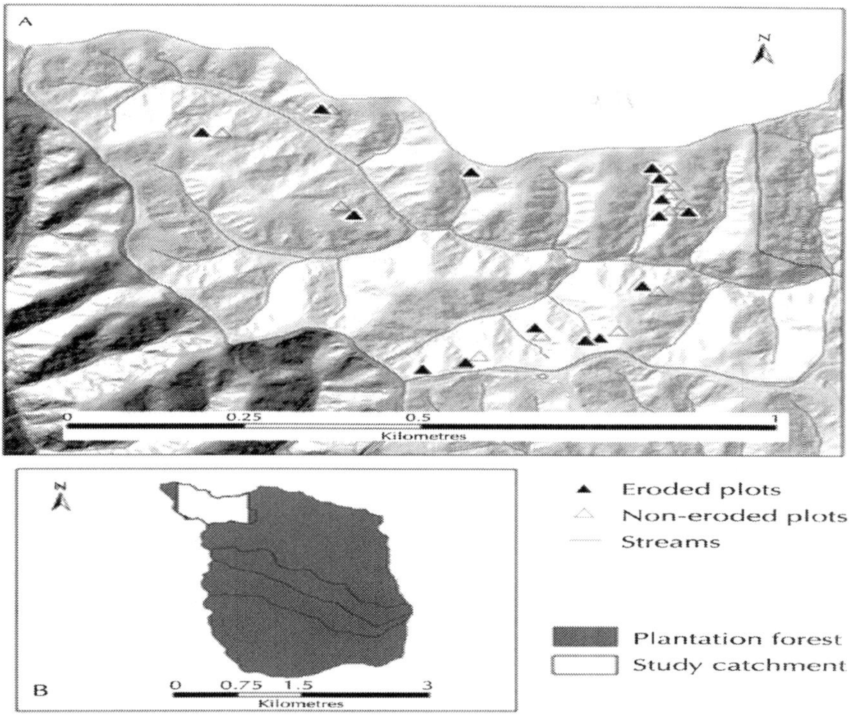

Figure 2: (A)Eroded and non-eroded paired plots forthe study area,and(B) location within the Pakuratahi forest catchment(in white).

The number of trees available per plot ranged from 5 to 10 for the calculation of mean basal area with a total of 209 measured trees across the plots. Plot sizes ranged from 90 to 127 m^2, slopes from 10 to 33°, and aspect from 2 to 358°. The plot locations were geo-referenced using a high-quality, global-navigation satellite system receiver (Trimble ProXRT). The raw data were differentially corrected using Trimble GPS Pathfinder Office Software (v5.3), a process which accesses the Land Information New Zealand (LINZ) network of base stations to improve survey-point accuracy.

Field Data Collection

Twenty soil samples per plot were collected for chemical analysis from 0 to 0.1 m depth using a small-diameter (23-mm) tube sampler. The samples were collected across each plot on a grid basis according to a stratified random pattern and were bulked for analysis on a per plot basis. Depth to impeding layer was assessed at each of the 20 sampling points using a 10-mm diameter metal rod. Soil cores (0–0.1 m depth) for bulk density measurements were collected at three random locations throughout each slip using stainless steel rings of 98 mm internal diameter. Soil sampling was carried out to a depth of 10 cm to accommodate sampling in stony soils. Six observations per plot were made using a Dutch auger to determine the thickness (depth) of the A horizon. In each plot, trees were measured for diameter-at-breast height (DBH – measured at 1.4 m from the ground) and tree height. Individual tree form was recorded for tree growth and tree health. The presence of faults was summed per plot and a two-tailed *t*-test for the average of two means was performed to evaluate the difference. Tree-form characteristics were multi-leader (or malformed), forked or dead top, and leaning sweep and toppled but still alive (Ellis and Hayes[1997]).

Soil Loss Measurements

The erosion scars used in this research were measured in the field. Only fourteen plot pairs were included in the final analysis because one erosion plot was washed out during an intense storm before the end of the trial. Depth and width measurements were taken at 5-m intervals along the length of each scar. Length measurements were adjusted for slope to determine the equivalent horizontal distance. Each scar was measured by suspending a stringline across the slip, first measuring the slip width and then its depth at the central position of the slip scar as set out in Page et al. ([1999]).

Sample Preparation and Nutrient Analysis

Chemical analyses of soil samples taken from paired non-eroded control plots were used to estimate nutrient loss from each slip. The rationale for using soil results from control plots was to simulate the antecedent soil profile that had subsequently been lost through erosion. Soils were sampled to a depth of 0.1 m and results averaged across all plots. Soil samples for chemical analysis were air dried at 20°C and passed through a 2-mm sieve (i.e. fine-earth fractions were analysed). Total carbon and total nitrogen were analysed by thermal combustion (Leco [2003]). Total phosphorus was measured by flow injection analysis (FIA) colorimetry after sequential Kjeldahl digestion (Blakemore et al. [1987]). Phosphorus availability was measured by FIA colorimetry after each of three sequential Bray-II extractions using 1:10 NH_4F/HCl (Bray and Kurtz[1945]; Skinner et al. [1991]). Soil pH was determined using an electronic probe and a 1:2.5 soil-to-water ratio with distilled water (1:2.5 dH_2O). Soil organic matter content was measured by loss on ignition at 550°C. Bulk density was determined from the oven-dry weight of bulked volumetric samples from each plot. Samples were air dried and sieved to < 2 mm. The coarse > 2-mm material was discarded. The < 2-mm fraction were oven dried at 105°C for 24 hours and then weighed again to determine oven-dry weight and fine fraction bulk density.

Statistical Analysis

Paired *t*-tests for sample means were used to test for significant differences in soil properties, and tree height and volume, between the eroded and non-eroded sites. Relationships between measured soil properties and tree volumes were determined using the Microsoft 2010 regression analysis tool. A linear regression analysis was performed using the 'least squares' method to fit a line through the ratio of slip data over non-slip data.

Prediction of Growth: Forecaster Modelling

Growth simulation software, Atlas Forecaster (West et al. [2012]) was used to estimate the economic return from trees measured in this trial. The Forecaster software enables the modelling of potential revenue. Forest mensuration data consisting of tree height, DBH, and tree form, were converted into stem lists and imported into GeoMaster (forest activity planning and management software: Atlas [2012]). PanPac Ltd provided additional silvicultural data including planting dates, stocking, pruning, and clearfell ages and the resulting analysis was based on a 22-year rotation (Table 1).

Table 1: Silvicultural regime for first-rotation planted forest

Date	Event	Stocking(stems ha-1)
1989	Planting	1200
1994	Pre-pruning	1190
1995	Pre-pruning	1190
1997	Pre-pruning	1190
1998	Post waste thinning	307
2012	Measurement	307
2014	Pre clear fell	298

Heaphy *et al.*

Heaphy *et al. New Zealand Journal of Forestry Science* 2014 44:24, doi: 10.1186/s40490-014-0024-5

Additional information required for the model included accurate site locations and elevation, which were identified using a Trimble global positioning system (GPS). Data were captured using the New Zealand Transverse Mercator (NZTM) mapping projection defined in terms of the New Zealand Geodetic Datum 2000 co-ordinates system. The Atlas Forecaster model was used to investigate plantation forest data using standard Ministry of Agriculture and Forestry log grades for the cutting strategy and default function models specific to the site (MAF [2011a]) (Table 2).

Table 2: Domestic log grades (MAF 2011a)

Log grade	Log status	Small end diameter(mm)	Maximum knot(mm)	Sweep class
P1	Pruned	400 +	0	1
P2	Pruned	300-399	0	1
S1	Unpruned	400 +	60	1
S2	Unpruned	300-399	60	1
S3	Pruned or unpruned	200-299	60	1
L1	Unpruned	400 +	140	1
L2	Unpruned	300-399	140	1

L3	Unpruned	200-299	140	1
Pulp	Unpruned	100	n/a	2

Heaphy *et al.*

Heaphy *et al. New Zealand Journal of Forestry Science* 2014 44:24, doi:10.1186/s40490-014-0024-5

Log grades were established, determining log volume and value (MAF [2011a]; [2011b]) (Table 3) for pruned trees, structural timber, utility grade timber, and pulp.

Table 3: Log grades and prices measured in Japanese Agricultural Standard cubic metres, free on board (loaded onto a ship at export port) as at Dec 2012 (MAF [2011b])

Generic log type & pricing point	Dec-12 quarter($)	12-quarter average($)a
EXPORT (NZ$ per Japanese Agricultural Standards m3free on board)b		
Pruned	144-190	168
Unpruned A grade	103-125	126
Unpruned K grade	90-121	116
Pulp	79-102	109
DOMESTIC (NZ$ per tonne delivered at mill)		
P1	122-149	138
P2	111-123	118
S1	95-104	99
S2	90-97	95
L1 and L2	77-96	88
S3 and L3	77-86	81
Pulp	48-53	51

[a]Average log prices over preceding three years.

[b]Log prices can be quoted at a number of different pricing points, such as "cost insurance freight" (or c.i.f., i.e., landed at an importing port), "free on board" (or f.o.b., i.e., loaded onto a ship at an export port), wharf gate, on truck, and at stump. Source: http://www.mpi.govt.nz/news-resources/statistics-forecasting/forestry/indicative-new-zealand-radiata-pine-log-prices.aspx webcite

Heaphy *et al.*

Heaphy *et al. New Zealand Journal of Forestry Science* 2014 44:24, doi:10.1186/s40490-014-0024-5

RESULTS

Soil Data

The total surface area of both the 14 measured eroded plots and 14 non-eroded plots was 0.25 ha for each. Erosion scars ranged in size from 9 to 15 m in width, 12 to 24 m in length, and 0.5 to 2.3 m in depth. Based on estimated soil volume from the slip measurements and measured bulk density, a total of around 415 tonnes of soil was estimated to have been displaced from the 14 eroded sites. The average losses of nutrients are outlined in Table 4. Values of measured soil properties were significantly lower in eroded plots than non-eroded plots with the exception of bulk density and the mean thickness of the A horizon (Table 5).

Table 4: Soil and soil nutrient loss estimated for 14 erosion plots in Pakuratahi (95%confidence interval in parentheses)

	Average per plot(n = 14)(kg)	[redacted]
Soil	296814 (±133265)	62737 – 870050
Total carbon	5864 (±2408)	1064 – 13920
Total nitrogen	438 (±183)	68 – 1131
Total phosphorus	63 (±26)	10 – 157

Heaphy *et al.*

Heaphy *et al. New Zealand Journal of Forestry Science* 2014 44:24, doi:10.1186/s40490-014-0024-5

Table 5: Mean soil properties across all plots using a paired sample two-tailed*t*-test

	Non-eroded	Eroded	% Difference	
Total carbon (%)	3.7	2.0	47	***
Total nitrogen (%)	0.3	0.1	52	***
C:N ratio	12.9	14.0	8	*
Total phosphorus (mg kg−1)	371.0	215.0	43	***
Soil organic matter (%)	8.9	5.7	36	***
Bray-II P 1st extraction (mg kg-1)	52.0	33.0	36	*
Bray-II P 2nd extraction (mg kg-1)	27.0	16.0	41	*
Bray-II P 3rd extraction (mg kg-1)	16.0	10.0	38	*
Soil pH	5.6	5.5	3	*
Mean soil depth (m)	0.4	0.3	16	*
Mean A horizon thickness (m)	0.2	0.2	12	ns
Bulk density (g m−3)	1.1	1.0	5	ns

[a]Probability: ***$p=0.001$; ** $p=0.01$; *$p=0.05$; ns = not significant.

Heaphy *et al.*

Heaphy *et al. New Zealand Journal of Forestry Science* 2014 44:24, doi:10.1186/s40490-014-0024-5

Three sequential Bray-II extractions (Bray and Kurtz [1945]; Skinner et al. [1991]) were carried out to provide an indication of available P in soil solutions and also long-term P supply. Although the second and third extracts each contained lower amounts of P than the first extract, the amounts were still substantial. These results demonstrate the capacity the Pakuratahi soils have to supply P over a period of time in both eroded and non-eroded plots.

Tree Growth and Form

Because of the small plot size (constrained by erosion scar size), average tree size within the study plots was used as a measure of productivity. A significantly larger mean DBH ($p=0.001$) of 9% and a volume of 14% ($p=0.01$) were recorded in trees from non-eroded plots compared with these parameters for trees measured in eroded plots. A two tailed t-test showed no difference in tree height and form with less than 1% difference between paired plots (Table 6).

Table 6: Mean tree properties across all plots using a paired sample two-tailed*t*-test

Characteristic	Non-eroded	Eroded	Probabilitya
DBH	493 mm	536 mm	***
Height	40 m	36 m	ns
Form	1.1	1.0	ns

[a]Probability: ***$p=0.001$; ns = not significant.

Heaphy *et al.*

Heaphy *et al.* *New Zealand Journal of Forestry Science* 2014 44:24 doi:10.1186/s40490-014-0024-5

Soil Properties and Tree Volume

Ratios of eroded plot data to non-eroded plot data were used for a linear regression analysis to determine relationships between tree volume and measured soil properties (Table 7). Individually, the soil properties that were most strongly related to volume increment were total carbon, nitrogen and phosphorus, and soil organic matter. Multilinear regression of soil properties combined did not result in any significant relationships.

Table 7: Summary of linear regression analysis between the ratios of eroded to non-eroded soil properties and*Pinus radiata*tree volume

Ratio of eroded to non-eroded soil data	Slope	pvalue	Probabilitya
Total carbon	0.2	0.0154	**
Total nitrogen	0.2	0.0136	**
Soil organic matter	0.2	0.0183	**
Total phosphorus	0.2	0.0324	*
Carbon to nitrogen ratio	−0.4	0.2512	ns
Soil pH	−1.7	0.2122	ns
Mean soil depth	−0.3	0.2113	ns
Mean A horizon thickness	0.1	0.1537	ns
Bulk density	−0.1	0.8678	ns

[a]Probability: $**p = 0.01$; $*p = 0.05$; ns = not significant.

Heaphy *et al.*

Heaphy *et al. New Zealand Journal of Forestry Science* 2014 44:24, doi:10.1186/s40490-014-0024-5

Effect of Erosion on Recoverable Volume and Log Value

The economic returns from a planted forest can vary depending on the quality of logs harvested. The calculated total recoverable log volume was 10% lower in eroded plots than in non-eroded plots. Eroded plots yielded 16% less volume from high-quality pruned logs, which accounted for a reduction in revenue of around $4,000 per hectare at eroded sites. However, eroded plots had higher volumes of lower grade sawlogs (S2, S3, and L3) than in non-eroded plots (Figure 3). A reduction in total recoverable revenue was estimated at $7,500 per hectare on eroded sites. Total recoverable volume (TRV) estimated (for a 25-year rotation) for *Pinus radiata* growing on the eroded sites was valued at $68,494, which is 9% lower than the estimated TRV from trees growing on non-eroded plots ($75,989) (Figure 4).

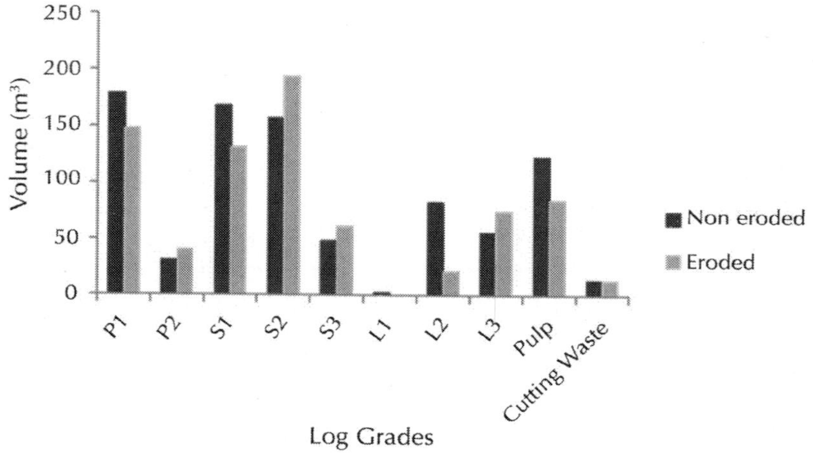

Figure 3: Total log volumes for log grades at eroded and non-eroded plots estimated for a 25-year-old*P. radiata*rotation. Pruned sawlogs (P): appearance-grade timber; small-branch sawlogs (S): structural timber; large-branch sawlogs (L): industrial-appearance timber; residual logs: pulp and panel products (Pine Manufacturers [2013]).

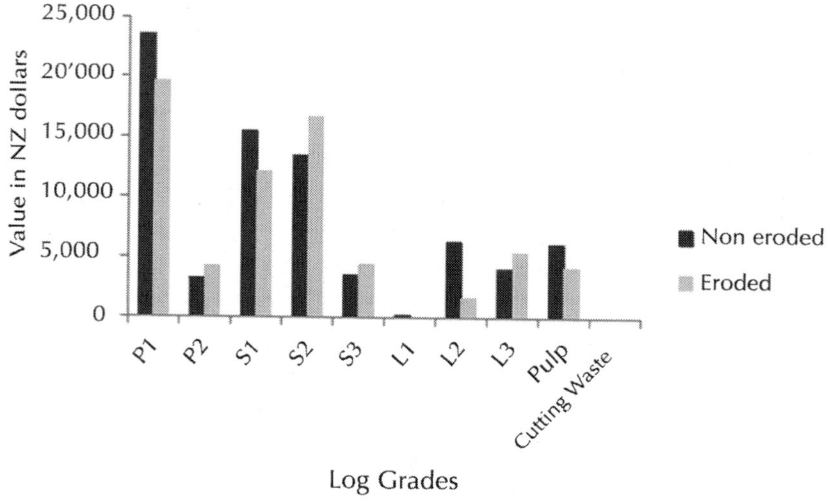

Figure 4: Log values for log grades in eroded and non-eroded plots.P = pruned trees: high-quality straight logs; S = unpruned (structural timber); L = unpruned (utility timber); pulp for paper; cutting waste which is left onsite.

DISCUSSION

Nutrients required by growing trees are provided by the soil and are immobilised in tree tissues in the early stages of tree growth and development (Attiwill and Adams [1993]). Trees become more dependent on nutrient cycling within the forest ecosystem when canopy closure is reached, receiving inputs from breakdown of forest litter and the atmosphere (Carnol and Bazgir [2013]). Hence trees planted in eroded sites may be disadvantaged in the early stages of growth because of the absence of topsoil and forest litter. In the study, trees growing in eroded soils were found to be significantly smaller in volume compared with trees in non-eroded soils. These trees were a first-rotation crop, planted in scrubby pastureland and aged 21 years at the time they were measured. Analysis of soils from the trial plots showed significantly fewer nutrients in eroded plots compared with non-eroded plots. In addition, regression analysis indicated a significant correlation between reduced amounts of C, N, P, and soil organic matter in eroded plots, and smaller tree volume. Although no one variable stands out as the main contributor to a lesser tree volume, each deficiency appears to have made an impact. The relationship between tree volume and P was not as strong as the relationships between tree volume and C, N, and organic matter.

Measured soil carbon has been found to decrease following afforestation of grasslands with *P. radiata* (Parfitt and Ross [2011]) but increases over time as pine litter accumulates, resulting in a higher soil total C content (Davis and Condron [2002]; Paul et al. [2002]; Laganière et al. [2010]). An increase in soil nutrients over time due to litter accumulation is quite likely to occur at Pakuratahi, but it is not certain how much of an impact litter has had on eroded sites because the ages of the erosion scars are difficult to determine. At the same time, inadequate levels of inorganic nutrients such as nitrogen and phosphorus in soils of existing erosion scars will have had an effect on tree growth and therefore on the rate and amount of organic matter accumulation.

Phosphorus deficiency is a common limiting factor in plantation forestry even though *P. radiata* has the advantage over some other plant species of being able to access soil organic P because of an association between radiata pine roots and ectomycorrhizal fungi (Davis and Lang

[1991]; Davis et al. [1996]). In the current study, soil total P levels were found to be significantly lower in eroded plots compared with levels in non-eroded plots. In addition, the concentrations of P in sequential Bray-II P extractions of soil from eroded plots were below the 12 mg kg^{-1} level recommended for seedling establishment (Ballard [1970]). This result suggests that P concentrations in eroded plots were possibly limiting growth of *P. radiata* in the long term. This view is supported by the values of the initial Bray-II extractions of eroded soils which were nearly half the values of soils in non-eroded plots, thereby suggesting that the difference in P supply is associated with the difference in tree volume.

The positive relationship between nitrogen and tree volume agrees with the results from previous studies that examined tree volume response to nitrogenous fertilisers. An increase in volume increment of around 8 m^3 ha^{-1} yr^{-1} following nitrogen application was found in research carried out on New Zealand soils from the central North Island in planted *P. radiata* stands (Mead and Gadgil [1978]) Responses to N and P fertilisers on less fertile soils in Nelson were up to 17 m^3 ha^{-1} yr^{-1} (Mead & Gadgil [1978]) In South Australia, near Mt Gambier, nitrogen fertiliser applications resulted in increased basal area increment after two years when applied prior to thinning (Carlyle [1995]).

The 10% decrease in total net volume forecast for trees in eroded plots compared to that for trees in non-eroded plots is comparable to earlier research[b] carried out in Pakuratahi. Forest productivity was examined in a first-rotation *P. radiata* plantation at 20 paired plots on known eroded and adjacent control (non-eroded) land under an earlier first-rotation regime of planted forest in the Pakuratahi catchment. Their study found a 26% decrease in stem volume, with a lower yield of more valuable pruned logs. In British Columbia, Canada, logs from non-eroded land produced three times the volume of wood compared with volumes of logs removed from areas eroded by landslides aged 30 to 59 years (Smith et al. [1986]). Logs from eroded sites where tree age averaged 85 years were around one half the volume of normal second-growth stands of the same age.

[b]The 300 Index defines the stem volume mean annual increment (MAI) at age 30 years for a reference regime of 300 stems ha^{-1} (Kimberley et al. [2005]).

At Pakuratahi, the forecasted revenue was $7,500 per hectare less for total recoverable volume (TRV) and $4,000 less per hectare for high-quality pruned logs from eroded plots. A reduction of revenue of this size may not be sufficient to warrant active management changes such as fertilisation of eroded land. However, the Pakuratahi planted forest is considered to be a high-production forest with a 300 index[b] rating ranging from 30 to 40 (Palmer et al. [2010]). The east coast climate is highly variable with summer droughts and high intensity rainstorms from northern tropical cyclones and cold southerlies from Antarctica (Salinger and Mullan [1999]). Hill-country forest land is vulnerable to intense rainstorms, particularly post-harvest when the soils are exposed. Site-specific management studies and economic analysis could determine whether fertilisation will provide an economic advantage in improving long-term productivity.

Following extreme weather events, such as Cyclone Bola (1988), there has been a move toward the afforestation of potentially unstable land with permanent and production forests, particularly in the worst-hit regions such as the east coast of the North Island. Government incentives supporting afforestation in New Zealand tend to focus on mitigating the effects of greenhouse gases such as carbon. However, the results presented here have shown that there is a cost to the land owner/manager for growing trees on formerly eroded land because of resultant smaller log volumes. The economic value of erosion prevention as a result of production forestry is not currently recognised in the returns from forest products in New Zealand. Nevertheless, the annual cost of erosion to New Zealand has been estimated to be around $200 million (Dymond et al.[2012]).

A framework to capture the natural capital and ecosystem services provided by forestry in New Zealand would help to identify key ecosystem services that are not currently recognised economically, such as planting of *P. radiata* for erosion control (Jones et al. [2008]). A recent framework developed for dairy farm soils identified a greater value from regulating than provisioning types of ecosystem services (Dominati et al. [2011]). In forestry only the provisioning services of log production are valued. Regulating services such as slope stability and erosion control (Phillips et al. [1989]) provision of habitat (Pawson et al. [2008]; Bremer and Farley[2010]) flood control (Serengil et al. [2011]) and the filtering of contaminants and nutrients near waterways (Dymond et al. [2010]) are yet to be incorporated into the benefits economically recognised in production forestry. If these services attracted financial benefits, land managers might be encouraged to explore alternative management methods to promote sustainable forestry and long-term site productivity.

CONCLUSIONS

Mass movement erosion in the Pakuratahi planted forest has resulted in a series of slips with associated erosion scars throughout the hilly and steepland catchment. Significantly lower levels of soil nutrients were detected in eroded soils than in non-eroded soils. This resulted in 47% less total carbon, 52% less total nitrogen, 43% less total phosphorus, and 36% less soil organic matter in the eroded topsoils compared with the non-eroded soils. These decreased levels of soil nutrients and soil organic matter significantly impacted on trees in eroded plots, resulting in a 10% decrease in volume compared with that of trees in non-eroded plots. The forecast value of these smaller logs means that trees grown in eroded sites are estimated to return a total recoverable volume of $7,500 per hectare less than that of trees growing in non-eroded sites. The loss in revenue is borne by the land or forest owner but could be offset if the ecosystem service of avoided erosion were recognised financially. If the costs and mitigation of the erosion-related loss in productivity prove to be too much for future forestry operations in hill country, government financial incentives may be warranted as a viable economic option to sustain the regulating services provided by forestry.

AUTHORS' CONTRIBUTIONS

DJL contributed to trial design, supervised thesis underpinning project, co-wrote paper. DP contributed to research conceptual design, supervised thesis underpinning project, co-wrote paper. HJ Assisted with project development, supervised thesis underpinning project, co-wrote paper. GG assisted with data analyses and interpretation, supervised thesis underpinning project, co-wrote paper. GO assisted with trial design, plot establishment, slip descriptive measurements, reviewed paper and assisted with rewrite in response to refereeing comments. SP assisted with trial design, plot establishment, technical field data collection and sample preparation for analysis. All authors read and approved the final manuscript.

ACKNOWLEDGEMENTS

This research was supported by the New Zealand Ministry for Science and Innovation through the Future Forest Research Programme "Protecting and Enhancing the Environment through Forestry" (Contract C04X0806). We acknowledge the assistance provided by PanPac Forests Ltd, and particularly Brett Gilmore and Brian Garnett who provided access to the forest and logistical support. We also thank Scion colleagues for invaluable support: Loretta Garret, Jason Bennett, Mark Kimberley, Peter Beets, Rod Brownlie, Duncan Harrison, Ruth Falshaw, Veritec Laboratory, and the Scion Knowledge Centre. Peter Clinton and Ruth Falshaw furnished valuable editorial comments. Additional assistance was provided by former University of Waikato technician Craig Hosking, and Kyle Bland of GNS Science, Wellington.

1. REFERENCES
2. (2012). Geomaster: Forest activity planning and management.
3. Attiwill, PM, & Adams, MA (1993). Nutrient cycling in forests. *New Phytologist, 124*(4), 561–582.
4. Ballard, R (1970). The phosphate status of the soils of Riverhead Forest in relation to growth of radiata pine. *New Zealand Journal of Forestry, 15*(1), 88–99.
5. Barry, LE, Yao, R, Harrison, DR, Paragahawewa, U, Pannell, DJ (2014). Enhancing ecosystem services through afforestation: How policy can help. *Land Use Policy, 39*, 135–145.
6. Blakemore, LC, Searle, PL, Daly, BK (1987). Methods for chemical analysis of soils. New Zealand Soil Bureau Scientific, Report 80. New Zealand Soil Bureau, Wellington.
7. Blaschke, PM, Trustrum, NA, Hicks, DL (2000). Impacts of mass movement erosion on land productivity: A review. *Progress in Physical Geography, 24*(1), 21–52.
8. Bray, RH, & Kurtz, LT (1945). Determination of total, organic and available forms of phosphorus in soils. *Soil Science, 59*, 39–45.

9. Bremer, LL, & Farley, KA (2010). Does plantation forestry restore biodiversity or create green deserts? A synthesis of the effects of land-use transitions on plant species richness. *Biodiversity and Conservation*, *19*(14), 3893–3915.
10. Carlyle, JC (1995). Nutrient management in a *Pinus radiata* plantation after thinning: the effect of nitrogen fertilizer on soil nitrogen fluxes and tree growth. *Canadian Journal of Forest Research*,*25*(10), 1673–1683.
11. Carnol, M, & Bazgir, M (2013). Nutrient return to the forest floor through litter and throughfall under 7 forest species after conversion from Norway spruce. *Forest Ecology and Management*,*309*, 66–75.
12. Crozier, MJ (1986). Landslides: causes, consequences and environment. Croom Helm, New Hampshire, USA.
13. Davis, MR, & Condron, LM (2002). Impact of afforestation on soil carbon in. *New Zealand: A review of paired site studies Australian Journal of Soil Research*, *40*(4), 675–690.
14. Davis, MR, & Lang, MH (1991). Increased nutrient availability in topsoils under conifers in the South Island high country. *New Zealand Journal of Forestry Science*, *21*, 165–179.
15. Davis, MR, Grace, LJ, Horrell, RF (1996). Conifer establishment in South Island high country: Influence of mycorrhizal inoculation, competition removal, fertiliser application, and animal exclusion during seedling establishment. *New Zealand Journal of Forestry Science*, *26*(3), 380–394.
16. Dominati, E, Mackay, AD, Green, S, Patterson, M (2011). The value of soil services for nutrient management. Adding to the knowledge base for the nutrient manager. 24th Annual Fertiliser and Lime Research Centre Workshop (Vol. Occasional report No. 24, pp. 1–8). Massey University, Palmerston North, New Zealand.
17. Dymond, JR, Betts, HD, Schierlitz, CS (2010). An erosion model for evaluating regional land-use scenarios. *Environmental Modelling and Software*, *25*(3), 289–298.
18. Dymond, JR, Ausseil, A-GE, Ekanayake, JC, Kirschbaum, MUF (2012). Tradeoffs between soil, water, and carbon – A national scale analysis from New Zealand. *Journal of Environmental Management*, *95*(1), 124–131.

19. Ellis, JC, & Hayes, JD (1997). Field guide for sample plots in New Zealand forests. FRI Bulletin, No. 186. New Zealand Forest Research Institute, Rotorua, New Zealand.
20. Eyles, G, & Fahey, B (2006). The Pakuratahi Land Use Study. Council, Hawkes Bay Regional.
21. Francescato, V, Scotton, M, Zarin, DJ, Innes, JC, Bryant, DM (2001). Fifty years of natural revegetation on a landslide in Franconia Notch, New Hampshire, U.S.A. *Canadian Journal of Botany, 79*(12), 1477–1485.
22. Fransen, P, & Brownlie, R. (1995). Historical slip erosion in catchments under pasture and radiata pine forest, Hawke's Bay hill country. New Zealand Forestry, 29–33.
23. Glade, T (1998). Establishing the frequency and magnitude of landslide-triggering rainstorm events in New Zealand. *Environmental Geology, 35*(2–3), 160–174.
24. Haywick, DW, Lowe, DA, Beu, AG, Henderson, RA, Carter, RM (1991). Pliocene-Pleistocene (Nukumaruan) lithostratigraphy of the Tangoio block, and origin of sedimentary cyclicity, central Hawke's Bay, New Zealand. *New Zealand Journal of Geology & Geophysics, 34*(2), 213–225.
25. Hewitt, AE (2010). New Zealand Soil Classification. Manaaki Whenua Press, Lincoln, New Zealand.
26. Jones, J, Clough, P, Hock, BK, Phillips, C (2008). Economic costs of hill country erosion and benefits of mitigation in New Zealand: Review and recommendation of approach. Ministry of Agriculture and Forestry, Wellington.
27. Kimberley, M, West, GG, Dean, MG, Knowles, RL (2005). The 300 index - a volume productivity index for radiata pine. *New Zealand Journal of Forestry, 50*, 13–18.
28. Laganière, J, Angers, DA, Paré, D (2010). Carbon accumulation in agricultural soils after afforestation. *Global Change Biology, 16*, 439–453.
29. Lambert, MG, Trustrum, NA, Costall, DA (1984). Effect of soil slip erosion on seasonally dry Wairarapa hill pasture. *New Zealand Journal of Agricultural Research, 27*, 269–285.
30. Lambert, MG, Trustrum, NA, Costall, DA, Foote, AG (1993). Revegetation of landslide scars in Wairarapa hill country. *Proc NZ Grasslands Assn, 55*, 177–181.

31. (2003). Total/organic carbon and nitrogen in soils (LECO Corporation). Organic application note 203-821-165, St Joseph, MO.

32. Lee, JM, Bland, KJ, Townsend, DB, Kamp, PJJ (2011). Geology of the Hawke's Bay area, Institute of Geological and Nuclear Sciences 1:250,000 geological map 8.1 sheet + 93 p. GNS Science, Lower Hutt.

33. Lowe, DJ, Blaauw, M, Hogg, AG, Newnham, RM (2013). Ages of 24 widespread tephras erupted since 30,000 years ago in New Zealand, with re-evaluation of the timing and palaeoclimatic implications of the late-glacial cool episode recorded at Kaipo bog. *Quaternary Science Reviews,74*, 170–194.

34. Lynn, IH, Manderson, AK, Page, MJ, Harmsworth, GR, Eyles, GO, Douglas, GB, Mackay, AD, & Newsome, PJF. (2009). Land Use Capability Survey Handbook 3rdEdition. Hamilton: AgResearch; Lincoln: Landcare Research; Lower Hutt: GNS Science. MAF. (2011a). Accessed 1 June 2011. MAF. (2011b). . Accessed January 11 2011.

35. Mead, DJ, & Gadgil, R (1978). Fertiliser use in established radiata pine stands in New Zealand. *New Zealand Journal of Forestry Science, 8*, 105–134.

36. Newsome, PFJ, Wilde, RH, Willoughby, EJ (2008). Land resource information system spatial data layers. Land Resource information system spatial data layers (pp. 72). Landcare Research, Palmerston North.

37. (2010). Cliflo: The National Climate Database.

38. Page, MJ, Trustrum, NA, Dymond, JR (1994). Sediment budget to assess the geomorphic effect of a cyclonic storm, New Zealand. *Geomorphology, 9*(3), 169–188.

39. Page, MJ, Reid, LM, Lynn, IH (1999). Sediment production from Cyclone Bola landslides, Waipaoa catchment. *Journal of Hydrology New Zealand, 38*(2), 289–308.

40. Palmer, DJ, Watt, MS, Kimberley, MO, Hock, BK, Payn, TW, Lowe, DJ (2010). Mapping and explaining the productivity of *Pinus radiata* in New Zealand. *New Zealand Journal of Forestry,55*(1), 15–21.

41. Parfitt, RL, & Ross, DJ (2011). Long-term effects of afforestation with *Pinus radiata* on soil carbon, nitrogen, and pH: a case study. *Soil Research, 49,* 494–503.
42. Paul, KI, Polglase, PJ, Nyakuengama, JG, Khanna, PK (2002). Change in soil carbon following afforestation review Forest. *Ecology and Management, 168,* 241–257.
43. Pawson, SM, Brockerhoff, EG, Meenken, ED, Didham, RK (2008). Non-native plantation forests as alternative habitat for native forest beetles in a heavily modified landscape. *Biodiversity and Conservation, 17*(5), 1127–1148.
44. Phillips, C, Marden, M, Rowan, D (1989). Planning for forestry after Cyclone Bola, a comment. New Zealand Forestry, November.
45. (2013). New Zealand pine users guide: log quality and conversion.
46. Restrepo, C, Vitousek, P, Neville, P (2003). Landslides significantly alter land cover and the distribution of biomass: An example from the Ninole ridges of Hawai'i. *Plant Ecology, 166*(1), 131–143.
47. Rosser, BJ, & Ross, CW (2011). Recovery of pasture production and soil properties on soil slip scars in erodible siltstone hill country, Wairarapa, New Zealand. *New Zealand Journal of Agricultural Research, 54*(1), 23–44.
48. Salinger, MJ, & Mullan, AB (1999). New Zealand climate: temperature and precipitation variations and their links with atmospheric circulation 1930–1994. *International Journal of Climatology, 19,* 1049–1071.
49. Serengil, Y, Swank, WT, Riedel, MS, Vose, JM (2011). Conversion to pine: Changes in timing and magnitude of high and low flows. *Scandinavian Journal of Forest Research, 26*(6), 568–575.
50. Skinner, MF, Lowe, AT, Nicholson, GM, Prince, J In White RE, Currie LD (Eds.) (1991). Availability of phosphorus in New Zealand forest soils: a new approach with the Bray reagent. *Soil and Plant Testing for Nutrient Deficiencies and Toxicities* (pp. 143–147). Massey University Occasional Report No. 5, Massey University: Palmerston North.
51. Smith, RB, Commandeur, PR, Ryan, MW (1986). Soils, vegetation, and forest growth on landslides and surrounding logged and old-growth areas on the Queen Charlotte Islands. Land Management Report No. 41. Canadian Forestry Service, Victoria, B.C.

52. Thompson, CS (1987). The climate and weather of Hawke's Bay (Vol. 5, 115). New Zealand Meteorological Service, Wellington.
53. Trustrum, NA, Lambert, MG, Thomas, VJ (1983). The impact of soil erosion on hill country pasture production in New Zealand. Paper presented at the Second International Conference on Soil Erosion and Conservation. Honolulu, Hawaii.
54. West, GG, Moore, JR, Shula, RG, Harrington, JJ, Snook, J, Gordon, JA, Riordan, MP (2012). Forest management DSS development in New Zealand. Technical University, Zvolen, Slovakia.

Chapter 7

Envisioning the Sustainability of the Production of Short Rotation Coppice on Grassland

Christine Rösch[1], Cisco Aust[2], and Juliane Jörissen[1]

[1]Institute for Technology Assessment and Systems Analysis (ITAS), Karlsruhe Institute of Technology (KIT), 76021 Karlsruhe, Germany

[2]Federal Forest Research Institute Baden-Wuerttemberg (FVA), 79007 Freiburg, Germany

ABSTRACT

Background

Short rotation coppices (SRC) offer a chance to enhance sustainable biomass production and to enlarge domestic wood resources. On arable land, SRC are less competitive in many regions compared to cropping systems and thus farmers are interested to establish SRC on grassland. This change of land use is discussed controversially because of the associated effects on the carbon-sequestering ecosystem grassland. In this paper, land use changes will be analyzed in reference to sustainability objectives, including aspects of land use competition with food and energy production or the preservation of biodiversity.

Methods

The Integrative Sustainability Concept has been adapted and applied to identify the relevant sustainability objectives using a holistic approach in order to assess the potential for the change in sustainable land use towards an expanded supply of lignocellulose-based bioenergy.

Results

A set of sustainability criteria with a special focus on the change in land use from grassland to SRC has been developed and discussed. For selected criteria, adequate sustainability indicators have been chosen and applied to a case study of the state of Baden-Wuerttemberg in Germany. The indicators have been implemented by performing a spatial differentiated bottom-up assessment using GIS-based data. The results revealed that around 13% of the grassland could be converted into SRC without harming the considered sustainability criteria.

Conclusions

The exploitation of the assessed potential for the change in sustainable land use from grassland to SRC could contribute to decentralize renewable energy supply, local employment and income. With regard to the economic values otherwise lost-, it is recommended not to set grassland conservation as an area-wide imperative. Instead strategies to direct the establishment of SRC to the most suitable areas and technologies should be developed realizing that land use transformation will be performed in a minimal invasive way to ensure compliance with the environment and climate protection.

BACKGROUND

The European Union (EU) has acknowledged the ecological functions of grassland and introduced measures within the Common Agricultural Policy (CAP) for its protection against conversion into arable land as well as for preserving or enhancing its ecological quality. Cross compliance, linking the full receipt of direct payments for farmers to compliance with environmental and further standards, explicitly addresses the protection of grassland (Regulation (EC) nos. 796/2004 and 73/2009). EU member states must ensure that at the national scale, the ratio of the land under permanent grassland in relation to the total agricultural area may not decrease by more than 10% compared to the year 2003. Germany implemented precautionary measures before the 10% threshold has been reached. A need for authorization to convert grassland to arable land is required if the 5% level is exceeded.

Resulting from the introduction of cross compliance in Germany, the ratio of permanent grassland is declining. By 2010, the 5% threshold was exceeded in four federal states. The federal states level is the level on which this ratio is calculated in Germany [1]. As a result, for any further conversion of permanent grassland, authorization has to be sought and as a rule, grassland has to be created elsewhere as compensation. More incentives for grassland are provided at the regional level in order to preserve certain management practices of extensive grassland. For less favored areas, natural handicap payments are given to support grassland management even under adverse conditions. However, these incentives could not stop the further decline of grassland either.

The main driving forces for the decrease of grassland are not only the continuously shrinking cattle stock being the primary user of grassland [2], but also the German Renewable Energies Act which makes the production of maize as a feedstock for biogas plants profitable. Around

53% of the permanent grassland that has been converted to arable land is being used to cultivate maize [1]. This land use change can induce negative impacts on the environment because grassland provides several important ecological benefits including carbon storage, protection of soil from erosion, ground water formation, and habitat function. Besides, grassland is forming an essential part of the cultural landscapes serving as a basis for nature-related recreation and tourism.

Since July 2011, the federal state of Baden-Wuerttemberg has prohibited the conversion of grassland by law. In the EU, the absolute maintenance of grassland is under discussion and could become one of the pillars of the projected ecologization of the CAP. Such a regulation would prevent any type of changes in grassland use regardless of their impacts. The justifications for such an absolute prohibition of grassland conversion are the multifunctional environmental benefits of grassland. Thereby the focus is on the functions of grassland as a host of endangered wildlife and plants and as a protector of huge carbon deposits which would otherwise be released as greenhouse gases. In Baden-Wuerttemberg, around 6% of the permanent grassland covering 532,000 ha or 38% of the agricultural used land is extensively managed, e.g. as meadow with scattered fruit trees [3]. This area is of great importance for the preservation of biodiversity. Part of the remaining grassland could be used for alternative biomass production. Permanent crops such as Miscanthus or fast growing trees have higher biomass yields compared to grassland, and the biomass has a higher amount of lignocellulose and a lower amount of nutrients that makes them more advantageous for renewable energy supply.

Against this background, it is the aim of this paper to evaluate the changes in land use from grassland to short rotation coppice (SRC) by reference to the Integrative Sustainability Concept. The focus is on SRC because fast-growing trees could contribute to the national wood resources and help to release the pressure on forests to become more productive in satisfying the growing demand for wood as an energy carrier which has been doubled in Germany within the last 5 years. The reserves to expand the use of wood in public forests without impairing sustainable and multifunctional forestry are limited due to ecological constraints. In small-size private forests with still higher wood reserves, structural problems prevent the mobilization of residual forest wood [4].

In Germany and also in other European countries, the cultivation of SRC has been intensively discussed and explored. According to the EU premium rights and to the German Federal Forest Act, SRC set up on agricultural land do not fall within the definition of forest. Hence, SRC maintain the legal status of agricultural areas as long as the rotation period is no longer than 20 years. Meanwhile, in Germany, around 4,000 ha of SRC have been established mainly for research and demonstration purposes. Energy scenarios are assuming that large agricultural areas are usable for SRC [5,6]. The European Environment Agency estimates that between 824,000 ha today and around 3.4 million ha in 2030 could be available for biomass crop production in the EU-15 (Austria, Belgium, Denmark, Finland, France, Germany, Greece, Ireland, Italy, Luxembourg, the Netherlands, Portugal, Spain, Sweden and the United Kingdom). More than half of which is available as former grassland [7].

METHODS

The Integrative Sustainability Concept [8] provides a theoretically well-founded approach to operationalize the Leitbild of sustainable development and an operable analytical tool for sustainability analyses. It has been applied so far successfully in various research projects on the global, national and regional levels [9]. The substantial minimum requirements of the Integrated Sustainability Concept (see Table 1) were chosen to identify relevant criteria for the assessment of the change in land use from grassland to SRC.

Table 1: Objectives of the integrative sustainability concept [8]

Securing mankind's existence	Upholding society's productive potential	Keeping options for development and action open
Protection of human health	Sustainable use of renewable resources	Equal access to education, information, and an occupation

Securing the satisfaction of basic needs	Sustainable use of non-renewable resources	Participation in societal decision-making processes
Autonomous self-support	Sustainable use of the environment as a sink	Conservation of the cultural heritage and of cultural diversity
Just distribution of chances for using natural resources	Avoidance of unacceptable technical risks	Conservation of nature's cultural functions
Compensation of extreme differences in income and wealth	Sustainable development of real, human, and knowledge capital	Conservation of social resources

The adaptation of the Integrative Sustainability Concept to the context of land use changes requires the identification of the relevant objectives as well as the selection of appropriate sustainability goals and indicators to measure the progress in reaching these goals. The sustainability objectives which have been considered to be most relevant in the context of land use change to SRC are listed in Table 2.

Table 2: Sustainability objectives for changes in grassland use

Sustainability objectives	Explanation
Securing the satisfaction of basic human needs	Affordable provision with basic supplies such as food and energy and prevention of competition between the use of commodities
Autonomous self-support	Adequate and stable economic existence by income-producing employment through agricultural land use
Sustainable use of renewable resources	Warranty of the efficiency, reliability and replacement of renewable resources
Sustainable use of non-renewable resources	Substitution of non-renewable fossil energy carriers with bioenergy
Sustainable development of real, human, and knowledge capital	Priority for feedstock production to operate existing food processing and bioenergy plants

Conservation of nature's cultural functions	Preservation of cultural and natural landscapes of particular characteristic and beauty

The conversion of grassland to SRC affects the objective to secure the supply of food as a basic human need because grassland is the main feedstock for milk and cattle meat production. The ongoing grassland conversion and expanding urbanization contribute to the continuous loss of this food production resource. At the same time, transforming grassland to high-yielding SRC production sites is detrimental to food supply. There is disagreement about how significant the 'food versus fuel' debate is, but consensus exists that biomass use for food should be prior to energy production.

Farmers provide a strong economic and social base for rural development. Diversification of income sources such as the increasing involvement of farmers in renewable energy production (e.g. biogas plants) helps to ensure the autonomous self-support of people working in the agricultural sector.

Agriculture has strong impacts on the use of the renewable resources soil, water and biodiversity. For example, agricultural practices are the single greatest contributor to the increase in soil erosion rates. Increasing production of renewable energy from biomass contributes to the substitution of fossil fuels and the avoidance of carbon emissions and climate change.

Agricultural activities rely on different types of capital including the real (natural) capital - the soil resource, water from rainfall or other sources, the air, animals used for their labour and as a source of manure, the surrounding natural vegetation; the human capital - humans who supply labour, not only physical labour but also intellectual input for planning production strategies; the social capital - systems providing labour and marketing support as well as information related to agriculture and health services; the financial capital - markets for purchase and sale of goods and a credit system supplying funds to all levels of agricultural workers; and human-made capital - implements needed for agriculture, roads and means of transport and factories for processing of farm produce. At every level, an agricultural system depends 'on the value of services flowing from the total stock of' these five types of capital. (Table 2).

The choice of objectives is accomplished by scientific expertise and based on reflections with regard to criteria relevance, diversity, comprehensibility and data availability. Consequently, the set of criteria is neither complete nor perfectly consistent with the initial proposition nor fully developed in terms of the complexities of systemic interactions and cross impacts. The preliminary character of the set of sustainability criteria proposed in this paper is amplified by the fact that an agreement has to be brought about regarding the selection of sustainability criteria preferably via negotiations within society and politics. Such a trans disciplinary approach, however, would have gone beyond the scope of the research described in this paper. In the following section, the characteristic aspects of the sustainability objectives listed in Table 2 will be discussed.

Satisfaction of Basic Human Needs

The ongoing increase in agricultural yield has first led to overproduction of food in Germany and the EU and then to the implementation of set-aside programmers to restrict food production. The subsequent set-aside areas have been unlocked from the production ban for the purpose to grow crops for material or energy supply. In Germany, now more than 2.15 million ha or 18% of the agricultural land are used for the production of renewable resources. Growth of crop yields has slowed down considerably, and fears are expressed that the trend may not reverse [10]. The question is whether the lower growth potential is sufficient to meet the increased requirements. Climate change, furthermore, looms largely as a risk that would negatively affect the production potentials of agricultural resources.

The increased request for agricultural land by the concurrent production of food and energy is visible in rising prices for the purchase and lease of agricultural land respectively in the neighborhood of biogas plants [11]. The price lift for arable land is much higher than for grassland, indicating the restricted land use options for grassland compared to arable land. The boosters of the advancing demand for land in Germany are the economic incentives given by energy politics in order to reach the ambitious renewable energy goals.

Food production should be the prior land use because renewable energy can be supplied by other types of renewable resources, such as wind, sun, water or geothermal power. The human need for food however can exclusively be covered by biomass produced on land and water-based production systems. Concerning this ethical aspect of land use, grassland needed for livestock grazing respectively for cattle farming or feeding other ruminants, such as sheep, should not be used for establishing non-food crops such as SRC.

The question though is how much grassland is needed to accomplish food supply today and in the future in a world based on the division of labour and resource use. In Germany, grassland is characterized by a long-term decline without visible impact on the food supply. Between 1990 and 2006, the area of permanent grassland in Germany declined at an annual rate of 0.8%, while arable land only decreased at a rate of 0.05% per year [1]. The overall loss of farmland due to urbanization seems to have occurred mainly at the expense of grassland [12]. This development is the result of the decrease in dairy cows due to political and economic constraints (i.e. the imposition of milk quotas) and the progress in breeding and milk production which have subjected milk production to significant structural changes for many years [13]. The average annual milk production per cow in Germany has increased from 4,710 kg in 1990 to 7,080 kg in 2010 [14]. Similarly, there has been a reduction in the number of beef cattle following the Bovine spongiform encephalopathy crisis in the late 1990s and the reduction and eventual removal of specific beef production incentives. As progress in breeding and dairy cow milk production will continue, a further decline in the demand of grassland can be expected.

Besides progress in breeding and milk production, the demand of grassland is influenced by the spatial allocation of the dairy and cattle stock and also by the intensity of dairy cow milk production and feed production on grassland and arable land. These factors are strongly dependent on economics, which in turn is influenced by agricultural policies and incentives. The objective of the German Sustainability Strategy to increase the share of extensive organic farming from 5.9% in 2010 to 20% will increase the demand of grassland for roughage production [15].

Since grassland is just one source of animal feed, its competiveness as compared to maize is crucial for its application [13]. Until the 1970s, grassland, lucerne and clover had been the main feeding base for cattle farming. Today, these traditional forage plants have been replaced continuously by maize. This is due to the area-specific higher yield of maize in contrast to grassland. This configuration will not be changed in the future as long as the spread of pests such as the corn rootworm will not limit maize production.

The conclusion is that grassland can be replaced by crops with higher yields per hectare such as maize. However, importing feed like soy bean from Latin American countries instead of using domestic grassland for the supply of animal feed raises ethical issues.

Autonomous Self Support

In common with most high-income countries, the agriculture's contribution to the German gross domestic product and employment is low, although agriculture is the largest type of land use accounting for about 52% of the total area. The regional importance of the sector and its associated food and farming industries varies considerably. The job and income opportunities associated with agricultural land use are strongly influenced by government interventions in agricultural markets. They are likely to be the most critical drivers of agricultural land use change with significant influence on the landscape.

The German agricultural policy has primarily been determined by the European CAP covering the key agricultural commodities. The CAP was devised to increase agricultural productivity, stabilize markets and assure the availability of reasonably priced food, and to ensure a fair standard of living for all those involved in farming. These objectives were largely met by the support of internal prices and incomes, through direct market intervention, and border protection through tariffs and levies on food imports. Export subsidies were paid to offload otherwise uncompetitive EU surplus production onto world markets. The policy was successful in terms of increasing domestic food production, to the extent that the EU became the second largest exporter of food in the world. However, this policy resulted in higher consumer food prices within the EU and increasing of tax burdens on EU citizens. Its legitimacy as a means of supporting farm and rural incomes was eventually challenged by food exporting nations and the World Trade Organization.

The drive for food production between 1940 and 1984 had led to substantial areas of permanent pasture being ploughed and drained, usually with grant support. By 1984, surplus milk production led to the introduction of marketable milk quotas. Eight years later, in 1992, the MacSharry reforms of the CAP led to constraints being placed on the beef and sheep sector, with compensation payments which were subject to regional ceilings and maximum stocking rates. At the same time, compensation payments in the arable crop sector required the compulsory withdrawal (or setting-aside) of up to 10% of arable land from production.

The negative environmental effects of intensive farming and land use change including loss of habitats and wildlife, soil erosion and water pollution have lead to a change in agricultural politics. Today, payment for farmers is not linked anymore to a particular type of production, but does require 'cross-compliance', an adherence to EU environmental, food safety, animal welfare standards and regulations, and the need to keep farmland in good agricultural and environmental condition. This is defined at a national level and includes the maintenance of grassland.

For grassland different incentives are available on the national and regional level, but for SRC, the payments are limited. With respect to the on-farm economic competitiveness of SRC toward the classical use of agricultural land under German conditions, it might be profitable to establish SRC, but today this is mainly not the case [15-17]. The establishment of SRC is hampered by traditionalistic behaviour and risk aversion, too. SRC involve investment costs that are at least partly sunk or irreversible, and they are changing the farm's revenues and expenditures over a long period of time. Farmers would go for SRC if the present value of the investment returns exceeds the investment costs considerably or additional and long-lasting area payments for SRC are provided by policy and society [18]. Farmers do not want to allocate their land for long periods of time and rather prefer great flexibility in their planting decisions in order to rapidly react to changes in prices for agricultural commodities.

Sustainable use of Renewable Resources

The EU has acknowledged the various ecological functions of grassland for the preservation of renewable resources and its benefits in terms of soil, water and biodiversity protection. Hence, the EU has introduced measures within the CAP for its protection against conversion into arable land as well as for preserving or enhancing its ecological quality. Extensively cultivated grassland belongs to the most species-rich land use types in Europe [19] and, after being destroyed, may only recover very slowly [20]. The conversion of this type of grassland could lead to a loss of biodiversity and should thus be excluded from changes in land use.

In recent times, grassland has received considerable attention for its carbon sink capacity and high carbon stock compared to arable land [21]. In particular, grassland on organic soils store very high amounts of carbon [22, 23]. The conversion from permanent grassland to arable land is accompanied by a massive boost of mineralization due to decomposition of soil organic matter and consequently rising greenhouse gas (GHG) emissions as well as higher nutrient runoff and leaching to surface and ground water. Converted grassland may lose about 50% or more of its original soil carbon under arable management [24]. In the case of temperate grasslands which contain in average around 330 mg of carbon per hectare; this is a tremendous release of GHG [25]. In contrast, a conversion of arable land to grassland results in increased carbon storage, however, at a much slower pace and rarely reaching the former level. Compared to annual agricultural crops such as rapeseed or maize, fast-growing trees have significantly lower negative impacts on the environment [24, 27-29]. This is related to the extensive form of cultivation which can lead to an increase of carbon and humus in the soil and their low demand for nutrients and pesticides [30, 31]. Established SRC have a high economic threshold to pest damage and can be managed with few pesticide applications. Tolerating the presence of certain insects and other plants is, in turn, cited as the main reason why SRC, if appropriately located, has the potential to increase biodiversity in many farmland situations [32].

The establishment of SRC on grassland is less valuable for the protection of resources and the environment than on arable land. The deterioration of the ecological functions of grassland can be reduced by the non-plough tillage, minimal invasive establishment of SRC and the abandonment of returning the SRC plantation into grassland or arable land [33]. Hence, SRC can contribute to both the buildup of humus and improvement of the soil quality, in particular, on soils with low carbon stocks [34]. This assumption is eligible because fast-growing trees in contrast to annual non-woody energy crops do not require good carbon-rich soils to achieve economically promising yields. They can also be produced on marginal agricultural sites with a low natural productivity, like sandy or stony soils; if the average temperature enables long vegetation periods, the requirements by the species are respected and water supply is guaranteed due to access to groundwater resources or sufficient precipitation [30].

SRC can also be beneficial for the protection of biodiversity within sparsely wooded pastoral landscapes where they could provide wooded stepping stones for birds of wooded habitats [35,36] and create networks of biotopes and enhance biodiversity [37,38]. Positive effects for biodiversity can also be expected when SRC are established in riparian buffer stripes to ensure that the regulations of distance regarding the application of fertilizer and pesticides will be respected [30] and in emission stripes along transportation routes to shield the fields behind against pollutants [31]. In water regulation areas, SRC can contribute to retard peak flows, decrease surface runoff and utilize the soil as a buffer [28, 30]. Against this background, it can be concluded that even on grassland the establishment of SRC can be beneficial for the preservation of renewable resources and the environment. The areas where this is true and which are not reaching a certain critical value of natural protection and the preservation of ecosystem services [27,28], should be identified and made available for establishing SRC. The measure for planting and operating SRC has to be conducted environmentally sound and needs to fit into the general principles for saving both natural and cultural landscape [27-29, 39].

Sustainable Use of Non-Renewable Resources

Energy and carbon analysis of different power generation processes based on SRC demonstrated that these processes use comparatively less fossil fuel and avoid more greenhouse gas emission per unit of produced energy than conventional high-input bioenergy crops [40]. The energy ratio of SRC is within the range of 22 and 28 depending on both the assumptions made with regard to the production and transformation processes and on the fact which energy consumptions along the process chain are included [41,42]. SRC are well positioned among the best energy crops and can contribute to increase the share of bioenergy compared to the energetic use of grassland and thus the amount of fossil energy carriers that can be substituted.

Sustainable Development of Real, Human and Knowledge Capital

In times of financial crises, the sustainable use of capital is of increasing importance for a sustainable development. Investments in bioenergy plants, such as biogas plants, which have been made in the last years based on the fixed feed-in tariffs specified by the German Renewable Energy Sources Act (EEG), could be endangered by rising prices for biomass feedstock due to an increasing competition for land use. Among the means of using grassland biomass, biogas production currently is the most common practice in Germany. Surveys of agricultural biogas plants in Germany show that grass silage is used as a feedstock in more than 50% of the biogas plants and is the second most frequent crop feedstock after maize silage [43]. Particularly, in regions with high percentages of grassland, the relevance of grass silage as a biogas feedstock could be quite high. To deal with the capital as a limited financial resource in a sustainable way, it is defined in this paper that grassland used to produce grass silage for already existing or planned biogas plants should not be taken into consideration for the establishment of SRC.

Conservation of Nature's Cultural Functions

The functions of nature are mainly associated with its life-support and ecological services. The important and irreplaceable cultural functions which are performed by natural environment are rarely investigated. This is due to the challenge to determine indicators and values for these sociocultural functions and their values for the health and well-being of human societies. The importance of the information functions (health, recreation, amenity, education, heritage, etc.) for the quality and sustainability of human life, however, is undisputed. Despite their immaterial and often intangible nature, these functions provide many socioeconomic benefits which might be assessed through both qualitative and quantitative evaluation methodologies. Depending on the spatial embedding and the design of SRC plantation, they can either enrich or depreciate the cultural landscape. The impact on the landscape largely depends on the suitability of the proposed location and on the current character and sensitivity of the landscape. In general terms, lowland landscapes with high levels of tree and woodland cover and arable or mixed farming are most suitable for SRC. It is also important to remember that cropping is heavily mechanized and requires land suitable for mechanical operations: This excludes steep or boggy ground.

The impact of SRC on the landscape will increase if whole farms or substantial parts of farms are planted. There is a risk that the diversity of the landscape will be reduced and that it will appear saturated [44]. It is important to build diversity into large-scale planting by varying age structure and introducing open space so that the crop is subdivided at a scale that suits the particular landscape type. Some landscapes, such as parkland and historically designed landscapes may be unsuitable for the introduction of SRC on any significant scale. In areas well used by the public such as near public rights of way, access routes and roadsides, a more detailed design may be necessary to respect cultural functions.

Indicators for Sustainable Grassland Use Change

The sustainability targets can be operationalized by the application of a readily comprehensible set of sustainability indicators. This way, the quality of the dialogue about sustainable land use change from grassland to SRC can be enhanced. The selection of suitable indicators is based on the following criteria: relevance, directional safety, availability of data, comprehensibility, and practicability. The indicators chosen to identify potential grassland areas which could be used for a sustainable establishment of SRC are listed in Table 3.

Table 3: Sustainability indicators for identifying grassland suitable for SRC

Sustainability objective	Sustainability indicator
Securing the satisfaction of basic human needs	Grassland no longer needed for animal feed production

Sustainable use of renewable resources	Grassland with suitable natural conditions with regard to temperature and water availability for fast growing trees
	Grassland with no particular function for the protection of biodiversity or ground water
Sustainable development of real, human, and knowledge capital	Grassland without feedstock functions for biogas plants
Autonomous Self support	Employment and income by grassland use

The analysis of sustainability indicators in this paper is focusing on the most important indicators regarding their relevance to illustrate the sustainability objectives which are strongly affected by the conversion of grassland to SRC and considering the availability of data. In the following, these indicators will be applied in the state of Baden-Wuerttemberg. Besides their purpose for sustainability assessment, they can be used to enhance communication and discussion in society and politics on options for sustainable land use changes. Moreover, they can contribute to provide opportunities for creativity in planning of how to achieve sustainable development and support decision-makers.

RESULTS

As a case of application of the selected sustainability indicators the federal state of Baden-Wuerttemberg in Germany has been chosen due to its still high percentage of grassland and the fact that its cattle stock is continuously decreasing. The pattern of grassland use is spatially diverse because of the various natural conditions influenced by different combinations of climate, soils and topography. Major grassland areas can be found in the region of the Black Forest and the Swabian Alb where agricultural land use is difficult due to sparse soils, slopes, temperature and high precipitation.

Grassland Suitable for SRC

Suitable temperature levels and the sufficient availability of water are crucial cultivation factors, especially for poplar and willow grown as SRC [45, 46]. The water availability can be described by the relief-dependent (influenced by the exposition, slope and relief position) climatic water balance within the period from May to October and the available water capacity in the upper 1 m of the ground. On 62 polar and willow plantations in south-west Germany and in the northern part of France, the water availability in millimeter and the yields in $tons_{dm}$ after the first three-year rotation were measured. On the basis of this data, a correlation between the water availability and the achievable yields could be derived and expressed in an equation [47]. The water availability was classified in five water availability stages (<0, 0 to 100, 100 to 200, 200 to 300 and >300 mm) to better visualize the natural SRC cultivation potential in maps. These classes represent the suitability of a site for the cultivation of poplar or willow grown as SRC. The classes are sorted in ascending order from <0 to >300 mm in these categories: excluded, unfavorable, medium, favorable and very favorable. The achievable yields on wet and waterlogged sites were also measured and classified as favorable and unfavorable by a mean annual temperature of more than 8°C. To consider the influence of the temperature on the growth of SRC, the mean annual temperature was classified in three classes: <6.5°C, 6.5 to 8.0°C, and >8.0°C; and the SRC site classes were downgraded according to the tempe- rature class (see Table 4).

Table 4: Classification of location suitability for short rotation coppice according to water availability and mean annual temperature

Annual temperature (°C)	Water availability stages						
	<0 mm	0-100 mm	100-200 mm	200-300 mm	>300 mm	Wet	Water logged
<6.5	Excluded	Excluded	Excluded	Excluded	Excluded	Excluded	Excluded
6.5-8.0	Excluded	Excluded	Unfavourable	Medium	Favourable	Medium	Excluded
>8.0	Excluded	Unfavourable	Medium	Favourable	Very favourable	Favourable	Unfavourable

Rösch et al.
Rösch et al. Energy, Sustainability and Society 2013 3:7, doi:10.1186/2192-0567-3-7

Since average temperature and water availability have a strong influence on the land use efficiency and economic performance of SRC, these criteria have particularly been chosen to identify land which is suitable for the establishment of SRC in the research region Baden-Wuerttemberg. In applying these criteria, it can be concluded that around 12% (65,000 ha) of the grassland in Baden-Wuerttemberg shows very favourable and around 30% (160,000 ha) favourable conditions for the cultivation of SRC. In total, over 40% of the grassland can be rated as suitable for fast-growing poplars and willows (see Figure 1).

Figure 1: Suitability of grassland area in Baden-Wuerttemberg for SRC with regard to site conditions [47].

The percentage of suitable land for poplar and willow SRC within the respective municipalities is considerably high in the south-east of Baden-Wuerttemberg (see Figure 1). Also, parts of the north-eastern mountain regions and the area between the Black Forest and the Upper Rhine Valley possess considerable land potentials.

Grassland without Feeding Functions

In Baden-Wuerttemberg, the number of cattle has decreased from around 1.9 million cattle in 1970 to around 1 million cattle in 2010. In the same period, the number of dairy cows has declined from 750,000 to 350,000. Compared to the German average in annual milk production per cow of around 7,080 kg, the average milk in Baden-Wuerttemberg is significantly lower at 6,579 kg [14]. Thus a further decline in the number of dairy cows can be expected due to progress in milk production and structural changes [2]. This decline is associated with a release of grassland not needed anymore for the production of cattle feed. Part of this so-called surplus grassland has been converted in the last years to arable land in order to produce maize. Agricultural policies and structural changes have shifted the milk production to regions with favourable natural production conditions. The nationwide trade of milk quota has further decreased the number of dairy cows in Baden-Wuerttemberg. Within 3 years, almost 5% of the total milk production has been translocated to the other states [48]. This correlates with a dismount of 15,000 cows or a surplus area of 7,000 ha grassland.

For the calculation of the surplus grassland, a method was developed to assess and allocate the surplus grassland on the basis of different statistical data on land use, livestock and agricultural yields in Baden-Württemberg as described in [49]. Further data needed for the calculation such as the energy content of grass silages were raised by a special two-step survey among the local agricultural offices. First, based on these data, the roughage demand of cattle, horses and sheep was calculated. Then the production volume of the different types of grassland (meadows, pasture) as well as other fodder crops was calculated in relation to the average yield and energy content on district level. Finally, the production volume was balanced with the estimated roughage demand of the livestock. For more information, see [49]. According to this method for calculating the amount of surplus grassland, it has been assessed that in Baden-Wuerttemberg, currently around 100,000 ha of the permanent grassland are not needed anymore for roughage production. The spatial distribution of this surplus grassland is quite different (see Figure 2). The main part of this surplus grassland which could be used for establishing SRC is located in the region of the Black Forest and the Swabian Alb.

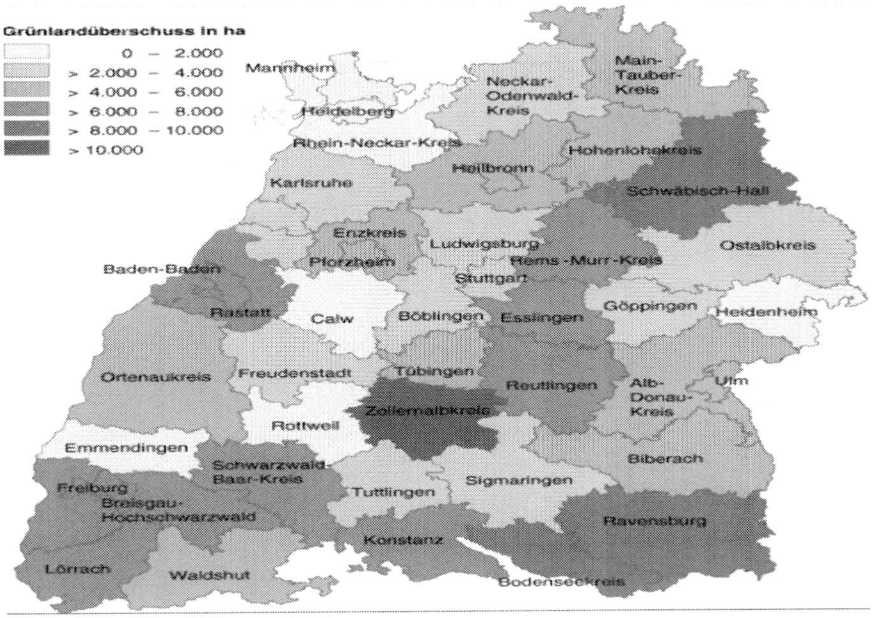

Figure 2: Grassland without a function for cattle feeding in Baden-Wuerttemberg.

At present, the exploitation of the potential of this grassland for SRC is limited by political frame conditions. Under the EU Cross Compliance Regulations within the Common Agricultural Policy, EU member states are required to ensure that land under permanent pasture is to be maintained. Under present EU legislation, there is a general derogation which allows some decreases provided that they are not significant - defining significant as losses which exceed 10% of a reference level - to be set by the member states themselves. In Germany, this level was set to 5% of the reference level to the correlation between grassland and arable land in 2003. In many states of Germany, the limit of 5% has already been reached. In the research region, only 24,000 ha or 2.38% of the referring grassland has been converted so far. Thus grassland could still be converted and used for fast-growing trees. However, legislation in Baden-Wuerttemberg has changed. Since December 2011, the conversion of grassland is prohibited mainly to prevent that more grassland is turned to maize fields for biogas plants. The establishment of SRC on grassland into SRC plantations is also generally excluded by law [50].

Grassland without Function for Biodiversity or Groundwater Quality

Environmental national and regional legislation in the areas of nature conservation and water and soil protection may limit the conversion of grassland in ecologically sensitive locations, such as nature protection areas and drinking water protection areas or flood plains, supporting both national and EU objectives. Grassland possessing a high variety of species is protected respectively by the FFH-guidelines [51], a specific EU legislation that classifies these areas as 'flora fauna habitats' (NATURA 2000 areas). NATURA 2000 is an ecological network of protected areas in the territory of the European Union based on the Habitats Directive and the Birds Directive, and protects around 18% of land in the EU countries. Hot spots of distribution for this type of land use in Baden-Wuerttemberg can be found in the mountain areas of the Swabian Alb and, more often, in the Black Forest. Grassland is also protected by the EU guidelines for bird protection, which correspond to 17.3% of the area of Baden-Wuerttemberg or about 620,000 ha of the NATURA 2000 areas. Amongst them, the proportion of grassland is about 110,000 ha, plus another 30,000 ha of fruit meadows. The protection status of permanent grassland in NATURA 2000 areas, being equivalent to 15.4% of the German land area, is not clear. In cases where they do not overlap with nature conservation areas, protected biotopes or other areas with defined restrictions on land management, grassland may be converted if a compatibility assessment is conducted featuring that no significant changes in land use are induced. Besides, land use changes are possible in these protection areas as long as this does not result in a significant deterioration or disturbance in relation to the protected habitats or species and it is ensured that habitat conditions would not be affected negatively [1]. This requirement is somewhat vague and apparently leaves some scope for conversion of grassland into arable land, particularly in the case of special protection areas designated under the Birds Directive.

About 25% of the grassland in Baden-Wuerttemberg is classified as 'extensively cultivated grassland' rich in species. In 8% of this area, the biodiversity is judged as very high [52] whereas 1% is classified to be a 'biotope grassland'. The agri-environmental programme of Baden-Wuerttemberg known as 'Marktentlastungs-

und Kulturlandschaftsausgleich' or MEKA II [53] acknowledged about 70,000 ha or 12% of the grassland areas as rich with species. This is just about half of the total extensively cultivated grassland classified this way. High amounts of this species-rich grassland are situated in the Swabian Mountains' districts of Zollernalb and Tuebingen. The preservation of biodiversity is regarded as superior sustainability goal than the production of renewable energy because this objective can be achieved by solar and wind energy too with less impact on biodiversity. Thus, the following categories of grassland are excluded from the establishment of SRC on grassland:

- Areas for the preservation of the diversity of species, officially nominated as Flora Fauna Habitats or Bird Protection Areas (NATURA 2000)
- Natural conservation areas, national parks, core zones of biosphere reserves, protected biotopes according to Section 30 of Federal Nature Conservation Act or BNatSchG,
- Permanent grassland cultivated extensively.

The high water demand of fast-growing poplars and willows can have an influence on the water availability in groundwater bodies. Thus, land areas with susceptible ground water regimes and crucial functions for water quality and functionality of water bodies are not regarded as suitable locations for sustainable cultivation of fast-growing trees. Figure 3 illustrates the spatial distribution of the areas for the protection of nature, landscape, biodiversity and water in Baden-Wuerttemberg.

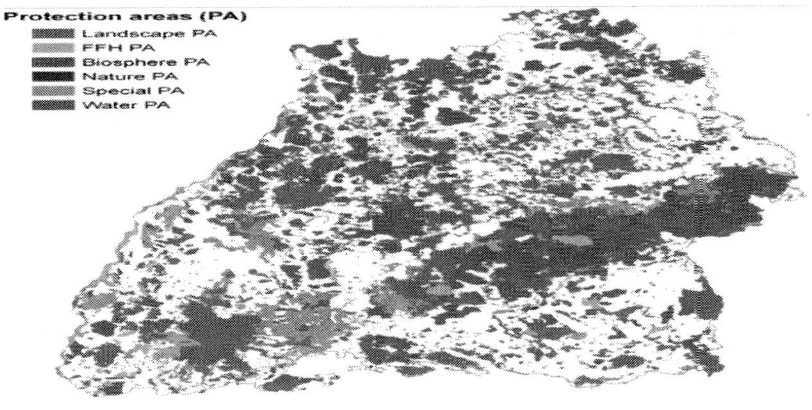

Figure 3: Spatial distribution of the areas for protection of nature, landscape, biodiversity and water in Baden-Wuerttemberg [47].

Figure 4 illustrates the areas on which the establishment of SRC is not possible or restricted and the area where fast-growing trees could be planted without impairment of biodiversity and water resources.

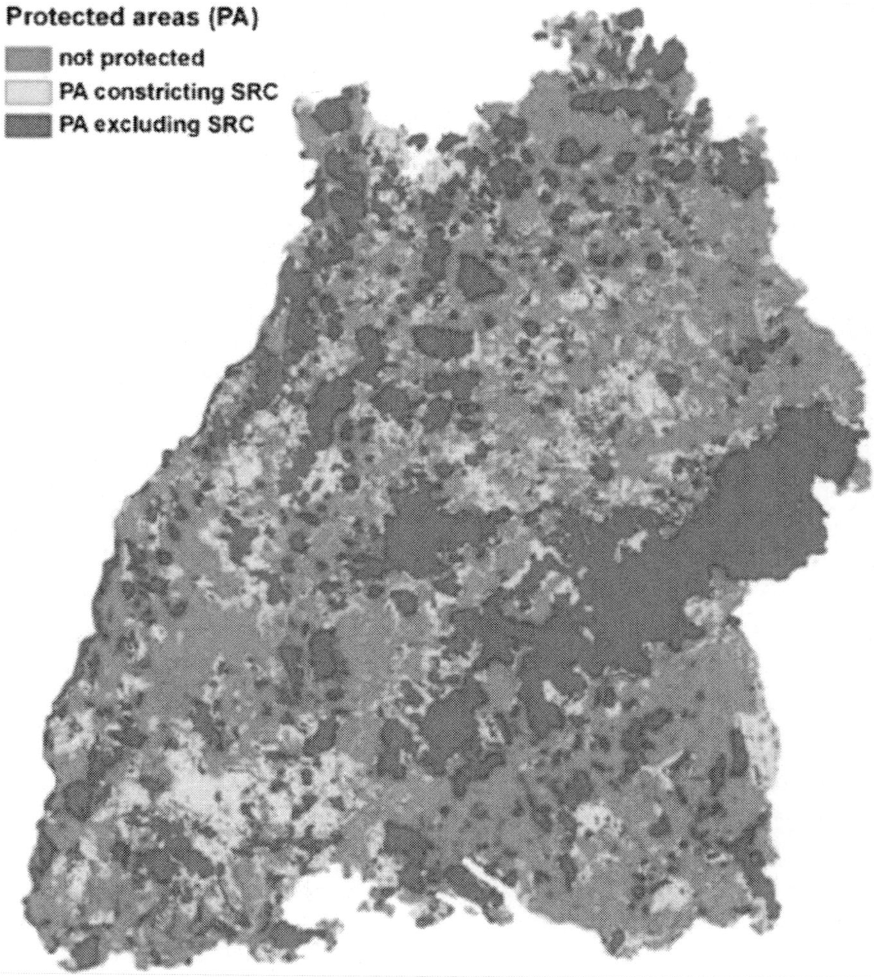

Figure 4: Locations where the cultivation of SRC is restricted due to environmental aspects [47].

Grassland without Biogas Feedstock Functions

Grassland provides grass silage which is a good feedstock for biogas plants that can easily be stored. Grass silage is applied in over 50% of the biogas plants and represents the second most important substrate type on a federal level providing 11% of the whole amount of biomass for biogas plants and allocating 12% of permanent grassland [43]. In Baden-Wuerttemberg, the average proportion of meadow grass differs between 9% in large plants >500 kW and 15% in small plants of 150 kW [54]. The distribution of biogas plants and the proportion of grass used for biogas plants are spatially different as shown in Figure 5. This is not only the influence of the spatial distribution of small and large-scale biogas plants but also due to the relation of grassland to arable land which is quite unequal in the various districts. In the south of the Black Forest (mainly in the districts Breisgau-Hochschwarzwald and Tuttlingen), grassland is the main source of feedstock for biogas production delivering more than a third of the biogas feedstock. In Baden-Wuerttemberg, the installed capacity of biogas plants has continuously increased until 2011 up to 256 MW of electricity. Assuming that in average 11% of the biogas feedstock supply for these plants derives from grassland; about 27,000 ha of grassland are needed to produce substrate for biogas plants. The increased use of grassland as a biogas feedstock resource could be one way to counter the widely expressed criticism of the significant expansion in maize cultivation for biogas plants. Despite its ecological advantages compared to maize, an enlargement of the share of grass feedstock is not assumed here because permanent grassland is less competitive in many regions compared to cropping systems [55]. The preference for growing silage maize for biogas production is based primarily on its high yields per hectare. In order to facilitate the use of grass, the Renewable Energy Act (EEG) 2012 introduces upper substrate limits for maize and cereal grain (ethical discussion) and different raw material payment classes (0, I, II). The intention is to expand the cultivation and use of grass and alternative substrates by providing a higher payment for the feedstocks.

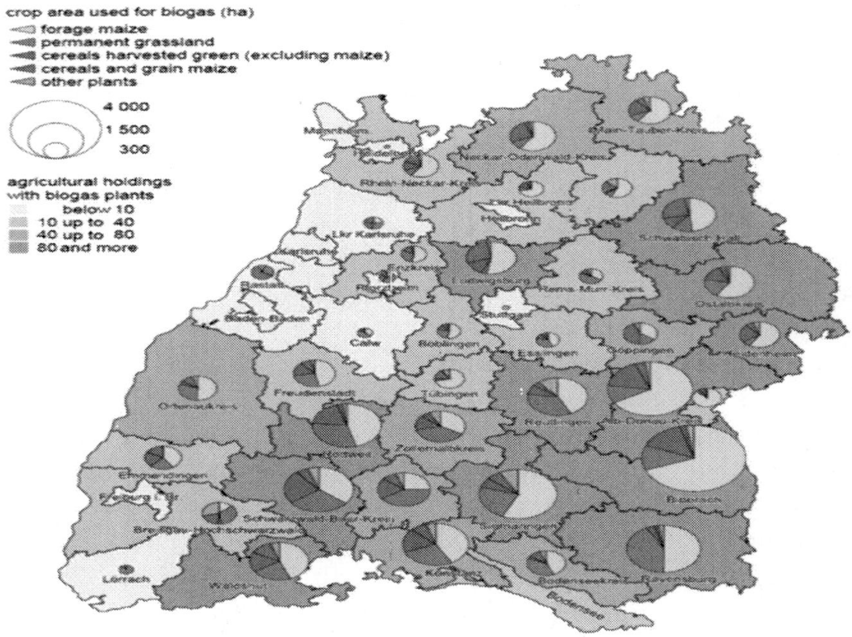

Figure 5: Grassland used for biogas production in the districts of Baden-Wuerttemberg [55].

This area of grassland assigned for biogas production is not considered to be disposable for establishing fast-growing trees in order to avoid feedstock competition. This prioritization is based on the sustainability goal not to jeopardize the sustainable use of real and human capital which has already been invested in local biogas plants and operation. Besides, the total area occupancy caused by biogas plants is already quite high in districts with large amounts of cattle livestock. Thus, negative impacts on the availability of land and land tenures have already been documented for such areas [56, 57]. Finally, only grassland which is not dedicated to biogas production should be considered for the establishment of SRC.

Employment and Income by SRC on Grassland

Milk and meat production based on dairy cows and cattle is the main source of employment and income from grassland. In Baden-Wuerttemberg, one quarter of the agricultural production value is generated on grassland [58]. As usual, the spatial distribution of this production value is differentiated. The main production areas are located in regions with favorable natural conditions for milk production such as the Allgaeu, Upper Swabia and the region from the Ostalb to Hohenlohe (see Figure 6). Over 30% of the total milk production is generated by two districts in the south-eastern part only, Ravensburg (464,100 t/a) and Biberach (226,000 t/a). However, also in these districts, grassland which is no longer used for milk production could be used for the establishment of SRC. The land in question is mainly marginal grassland with regard to the practicability for cultivation with large-sized efficient harvest machinery [2]. The establishment of SRC will result in a reduction of employment and income compared to grassland which is used as resource for milk and meat production due to a far less labour-intensive production chain. Yet, if SRC are grown on surplus grassland without functioning as dairy and cattle or biogas feedstock, a positive impact on employment and income for the local community can be realized. Hence, from the view of sustainable development, only the grassland area identified in the preceding chapters should be applied for the establishment of SRC.

Figure 6: Suitability of grassland for the cultivation of SRC in Baden-Wuerttemberg with regard to sustainability criteria.

Grassland Suitable for SRC with Regard to Sustainability

The results achieved by the application of the previous sustainability criteria exhibit that in the research region, around 13% or 70,000 ha of grassland are suitable for the establishment of SRC. A main part of this grassland can be found in the south-eastern part of Baden-Wuerttemberg (see Figure 6).

DISCUSSION

The sustainability debate on land use change is often narrowed and reduced to the questions of compatibility with environmental and climatic conditions. However, land use change has to be evaluated against a much larger spectrum of sustainability criteria, including aspects of security of supply and economic development of rural areas. The adoption of the Integrative Sustainability Concept to analyse land use change from grassland to SRC has the big advantage that socio-economic and ecological aspects are regarded together and of equal rank. From the 15 principles of the Integrative Sustainability Concept, six principles have been regarded as relevant to describe the impacts of land use change. Aside from the sustainable use of renewable resources, the criteria referring to the satisfaction of basic human needs, the autonomous Self support of the population and the development of real, human and knowledge capital have been identified as highly relevant for identifying grassland which could be transformed to SRC in a sustainable manner. Even if these restrictions are taken seriously, there are considerable areas of grassland available for the establishment of SRC as turn outs by the application of the sustainability principles in the state of Baden-Wuerttemberg.

The impacts on the cultural landscape heritage and nature's cultural functions, respectively, with regard to tourism and recreation are also crucial for assessing the sustainability and also the acceptability of land use change. Empirical surveys are needed to investigate the perception of land use changes from grassland to SRC on the levels of individuals, stakeholders, main actors and politicians, and group modelling is required to learn about the arguments and feelings behind the willingness for or resistance against land use change. In doing so, the preference of rejection of SRC has to be raised in comparison to other land use change for decentralized biomass and energy production technologies like open space photovoltaic or wind power. However, this would have been beyond the scope of this work. From lessons learned about land use change due to biogas plants and their demand of feedstock, it can be assumed that in particular large-scale SRC plantations could disrupt the character of the landscape and lead to the denegation of SRC. Hence, it will be of great importance to design and implement SRC in a way that the cultural and recreational landscape as well as human landscape perception, cognition and values are not affected negatively.

Regardless of the potential for sustainable land use change from grassland to SRC, the legislation in the state of Baden-Wuerttemberg prohibits the conversion of grassland at all. This regulation is a reaction to the conversion of grassland into arable land which begins with tillage, a practice known as negatively affecting soil properties and ecosystem processes in grassland-dominated ecosystems and to contributing to climate change by releasing soil-bound carbon stocks. On the EU level, the conversion of grassland is prohibited also but so far this rule is only applicable for land with high carbon stocks and high biodiversity which will be used afterwards for biofuel production according to the EU Renewable Energy Directive. These regulations are designed mainly for the protection of biodiversity; soil and climate provoke questions about the future of grassland which is and will be released from their traditional functions for roughage production and play no vital role for nature conservation purposes. An increase in the intensity of dairy farming and in biogas production as well as agricultural changes can lead to a continuing conversion of grassland, in cases when grassland is rented by other farms. Grassland without any production or nature conservation functions will no longer contribute to the farmer's income and the added value of the region but has to be maintained instead with public funds.

The ban on the change of grassland has not only consequences on local employment and income, but also on the decentralized supply and use of renewable and storable energy. Moreover, public funding of research on SRC in general and the development of methods to establish SRC are needed in a way that soil functions and carbon stocks are largely preserved. However, in other German states, such as the

Saarland, the establishment of SRC even on grassland is subsidized by the government with up to 1,000 euros per hectare in order to combat climate change with local actions. The objective is to establish SRC on marginal land where competition from other crops is less severe, due to their lower economic return, and with less intensive cultivation systems closer to traditional forest management.

CONCLUSIONS

The potential for a sustainable change in the use of land from grassland to SRC in south-west of Germany is likely to be significant and can contribute to decentralized renewable energy supply, local employment, and income. The exploitation of this potential is currently hindered by the prohibition to transform grassland to other types of land use in Baden-Wuerttemberg. This decision is strongly influenced by ecological aspects and does not consider the chances provided by SRC as an attractive alternative for a more extensive farming type and as additional local source for added value in economic and ecological terms. It is recommended to protect grassland more stringently on ecologically sensitive sites and also to strengthen legal standards of nature and water protection, but with regard to the lost economic values, not to set grassland conservation as an area-wide imperative.

In recognition of the advantages of SRC for sustainable land use, continuous research and demonstration of SRC are needed for improving the knowledge about site suitability and yields under current and future climates and frame conditions. Currently, only limited information is available for understanding the bottom-up spatial constraints for SRC production and both the changes in carbon stocks and the release of greenhouse gases, which are strongly dependent on the location, and the procedure how to establish SRC on grassland. More research is also required on supply chains to comprehend how demand and resource location affect costs and uptake in different end uses, and therefore inform on infrastructure and technologies that will be required.

Wood is the main renewable energy source in Germany and the government has ambitious targets to further increase the provision of bioenergy. Although all energy crops contribute to energy and climate-related sustainability goals, SRC are more favourable than annual energy crops to reach the targets. Nevertheless, the area of SRC in

Germany is still very small and only slowly rising. This is mainly due to economic constraints and the lack of competiveness because of lower profit and return on investment compared with traditional crops. New and better incentives are needed for the establishment of SRC. Besides, more differentiated subsidies strategies based on spatial land use considerations have to be developed in order to achieve a better direct land use change. To avoid repeating the mistakes which have been made with other subsidies for energy crops and to ensure that the public money is achieving overall profit, sustainability criteria should be applied to allocate the SRC in locations where they can contribute best to nature and society.

By applying these criteria in the state of Baden-Wuerttemberg in a spatially differentiated bottom-up assessment, it was found that around 13% of the grassland might be available for SRC. Further research is needed not only in order to specify this preliminary allocation process and to tap the identified potential, but also in order to clarify whether SRC should mainly be established on agricultural land which cannot be used in a highly productive way due to diverse local disadvantages such as along small rivers, very wet areas or small fields. Strategies and measures need to be developed and implemented to direct the establishment of SRC to the most suitable areas. Moreover, it is necessary to integrate stakeholders, main actors and the general public into the process of identifying areas for SRC. A major challenge will be to gain acceptance for the changes in landscape which will be evoked by establishing SRC, as they can lead to negative impact on the cultural and recreational functions of the landscape.

AUTHORS' CONTRIBUTIONS

CR formulated the concept and drafted the manuscript. CA participated in the spatial potential assessment. JJ contributed to the 'Methods' section. All authors have read and approved the final manuscript.

REFERENCES

1. Nitsch H, Osterburg B, Roggendorf W, Laggner B (2012) Cross compliance and the protection of grassland - illustrative analyses

of land use transitions between permanent grassland and arable land in German regions. Land Use Policy 29(2):440-448.

2. Rösch C, Skarka J, Raab K, Stelzer V (2009) Energy production from grassland - Assessing the sustainability of different process chains under German conditions. Biomass Bioenergy 33:689-700
3. Hartman A (2012) Dauergrünland in Baden-Württemberg. Statistisches Monatsheft Baden-Württemberg 2(2012):33-37
4. Mantau U, Saal U, Prins K, Steierer F, Lidner M, Verkerk H, Eggers J, Leek N, Odlenbuerger J, Asikainen A, Anttila P (2010) EUwood - real potential for changes in growth and use of EU forests. Hamburg, Germany: Final report. p 160
5. Fritsche UR, Dehoust G, Jenseit W, Hünecke K, Rausch L, Schüler D, Wiegmann K, Heinz A, Hiebel M, Ising M, Kabasci S, Unger C, Thrän D, Fröhlich N, Scholwin F, Reinhardt G, Gärtner S, Patyk A, Baur F, Bemmann U, Groß B, Heib M, Ziegler C, Flake M, Schmehl M, Simon S (2004) Stoffstromanalyse zur nachhaltigen energetischen Nutzung von Biomasse. Öko-Institut Darmstadt: Endbericht.
6. Nitsch H, Osterburg B (2010) Grünland - Bestandsaufnahme und Handlungsoptionen: Erkenntnisse aus dem InVeKoS-Projekt. Experten Workshop Einsatz von Grünlandaufwuchs und Landschaftspflegematerial in. Jena: Biogasanlagen.
7. EEA (European Environment Agency) (2007) Estimating the environmentally compatible bioenergy potential from agriculture, European Environment Agency. Technical report No 12/2007. Available at http://www.eea.europa.eu/publications/technical_report_2007_12 webciteAccessed 12 Dec 2012
8. Kopfmüller J, Brandl V, Jörissen J, Paetau M, Banse G, Coenen R, Grunwald A (2001) Nachhaltige Entwicklung integrativ betrachtet. Regeln, Indikatoren, Berlin: Konstitutive Elemente.
9. Kopfmüller J (2006) Ein Konzept auf dem Prüfstand. Das integrative Nachhaltigkeitskonzept in der Forschungspraxis. Berlin: Edition Sigma.
10. Alexandratos N, Bruinsma J (2012) World agriculture towards 2030/2050, ESA working paper no. 12–03. Agricultural Development Economics Division of the Food and Agriculture

Organization of the United Nations. Available at http://www.fao.org/docrep/016/ap106e/ap106e.pdf

11. DBV (Deutscher Bauernverband) (2012) Situationsbericht 2012/13. Trends und Fakten zur Landwirtschaft. Available at http://www.bauernverband.de/situationsbericht-2013. Accessed 12 Dec 2012
12. Osterburg B, Laggner B, Nitsch H, Röder N, Roggendorf W (2010) Analysis of grassland conversion into arable land in Northwest Germany. Available at http://www.egf2011.at/files/pubs/350_osterburg.pdf. Accessed 12 Dec 2012
13. Osterburg B, Isermeyer F, Lassen B, Röder N (2010) Impact of economic and political drivers on grassland use in the EU. Grassland Sci Europe 15:14-28
14. LEL (Infodienst Landwirtschaft-Ernährung-Ländlicher Raum, Schwäbisch Gmünd) (2012) Agrarmärkte 2011/12 Milch. Available at http://landwirtschaft-bw.de/pb/MLR.Landwirtschaft,Lde/Startseite/Marktinformationen/Milch+_+Milchproduktewebcite. Accessed 12 Dec 2012
15. The Federal Government (2008) Progress Report 2008 on the National Strategy for Sustainable Development. For a Sustainable Germany. Available at http://www.bundesregierung.de/Content/EN/_Anlagen/2009-05-13-fortschrittsbericht-englisch-nachhaltigkeit.pdf?__blob=publicationFile
16. Bemmann A, Feger KH, Gerold D, Grosse W, Hartmann KU, Petzold R, Röhle H, Schweinle J, Steinke C (2007) Kurzumtriebsplantagen auf landwirtschaftlichen Flächen in der Region Grossenhain im Freistaat Sachsen. Forstarchiv 78(2007):95-101
17. Wagner P, Heinrich J, Kröber M, Schweinle J, Grosse W (2009) Ökonomische Bewertung von Kurzumtriebsplantagen und Einordnung der Holzerzeugung in die Anbaustruktur landwirtschaftlicher Unternehmen. In: Reeg T, Bemmann A, Konold W, Murach D, Spiecker H (eds) Anbau und Nutzung von Bäumen auf landwirtschaftlichen Flächen, Wiley-VCH Verlag: Weinheim. pp 135-145
18. Musshoff O (2012) Growing short rotation coppice on agricultural land in Germany: a real options approach. Biomass Bioenergy 41:73-85

19. Stoate C, Baldi A, Beja P, Boatman ND, Herzon I, von Doorn A, de Snoo GR, Rakosy L, Ramwell C (2009) Ecological impacts of early 21st century agricultural change in Europe - a review. J Environ Manage 91(2009):22-46
20. Kleijn D, Kohler F, Baldi A, Batary P, Conception ED, Clough Y, DM Gabriel D, Holzschuh A, Knop E, Kovacs A, Marshall EJP, Tscharntke T, Verhulst J (2009) On the relationship between farmland biodiversity and land-use intensity in Europe. Proceedings of the Royal Society 276:903-990
21. Robert T, Watson R, Noble I, Bert Bolin NH, Ravindranath , David J, Verardo , Dokken David J (2000) Land use, land-use change and forestry. Cambridge: Cambridge University Press. p 375
22. Freibauer A, Drösler M, Gensior A, Schulze ED (2009) Das potenzial von wäldern und mooren für den klimaschutz in deutschland und auf globaler ebene. Natur und Landschaft 84(1):20-25
23. Lal R (2008) Carbon sequestration. Philos Trans R Soc B 363(2008):815-830
24. Guo LB, Gifford RM (2002) Soil carbon stocks and land use change: a meta analysis. Glob Chang Biol 8:345-360
25. Schlesinger WH (1997) Biogeochemistry: an analysis of global change. San Diego: Academic.
26. Peters W, Schultze C (2010) Bioenergie und Naturschutz. Synergien fördern, Risiken vermeiden. Bundesamt für Naturschutz (BfN): Bonn. Available athttp://www.bfn.de/fileadmin/MDB/documents/themen/erneuerbareenergien/bfn_position_bioenergie_naturschutz.pdf
27. BUND (Bund für Umwelt und Naturschutz Deutschland) (2010) Kurzumtriebsplantagen für die Energieholzgewinnung - Chancen und Risiken. Positionspapier. Available athttp://www.bund.net/fileadmin/bundnet/publikationen/landwirtschaft/20100714_landwirtschaft_bund_position_55_KUP.pdf
28. NABU (Naturschutzbund Deutschland) (2008) Energieholzproduktion in der Landwirtschaft. Chancen und Risiken aus Sicht des Natur- und Umweltschutz. Available at http://www.nabu.de/imperia/md/content/nabude/energie/biomasse/nabu-studie_energieholz.pdf

29. SRU (Sachverständigenrat für Umweltfragen) (2007) Klimaschutz durch Biomasse. Stuttgart: Sondergutachten. Available athttp://www.umweltrat.de/cae/servlet/contentblob/467474/publicationFile/34339/2007_SG_Biomasse_Buch.pdf
30. Ledin S (1998) Environmental consequences when growing short rotation forests in Sweden. Biomass Bioenergy 15:49-55
31. Ranney JW, Mann LK (1994) Environmental considerations in energy crop production. Biomass Bioenergy 6:211-228
32. Sage RB, Cunningham M, Boatman N (2006) Birds in willow short-rotation coppice compared to other arable crops in central England and a review of bird census data from energy crops in the UK. Ibis. 148:184-197
33. Rossi A (2012) Good environmental practices in bioenergy feedstock production. Making bioenergy work for climate and food security. Environment and Natural Resources management Working Papers 49. Available at http://www.fao.org/docrep/015/i2596e/i2596e00.pdf
34. Pellegrino E, Di Bene C, Tozzini C, Bonari E (2011) Impact on soil quality of a 10-year-old short-rotation coppice poplar stand compared with intensive agricultural and uncultivated systems in a Mediterranean area. Agric Ecosyst Environ 140(1–2):245-254
35. Fry DA, Slater FM (2011) Early rotation short rotation willow coppice as a winter food resource for birds. Biomass Bioenergy 35(7):2545-2553
36. Goransson G (1994) Bird fauna of cultivated energy shrub forests at different heights. Biomass Bioenergy 6:49-52
37. Berg A (2002) Breeding birds in short-rotation coppices on farmland in central Sweden-the importance of Salix height and adjacent habitats. Agri Ecosyst and Env 90:265-276.
38. Sage RB, Tucker K (1997) Invertebrates in the canopy of willow and poplar short rotation coppices. Aspects Appl Biol 49:105-111
39. Sage RB (1998) Short rotation coppice for energy: towards ecological guidelines. Biomass Bioenergy 15(1):39-47
40. Lettens S, Muys B, Ceulemans R, Moons E, Garcia J, Coppin P (2003) Energy budget and greenhouse gas balance evaluation of sustainable coppice systems for electricity production. Biomass Bioenergy 24(3):179-197

41. Matthews R, Robinson R, Abbott A, Fearis N (1994) Modelling of carbon and energy budgets of wood fuel coppice systems. London, UK: Department of Trade and Industry.
42. Dubuisson X, Sintzoff I (1998) Energy and CO_2 balances in different power generation routes using wood fuel from short rotation coppice. Biomass Bioenergy 15(4–5):379-390
43. DBFZ (Deutsches BiomasseForschungsZentrum (2011) Monitoring zur Wirkung des ErneuerbareEnergien-Gesetz (EEG) auf die Entwicklung der Stromerzeugung aus Biomasse. Available athttp://www.dbfz.de/web/fileadmin/user_upload/Userupload_Neu/Stromerzeugung_aus_Biomasse_Zwischenbericht_Maerz_2011.pdf webcite
44. Bell S, McIntosh E (2001) Short rotation coppice in the landscape. UK. Available at http://www.forestry.gov.uk/PDF/fcgn2.pdf/$FILE/fcgn2.pdf
45. Petzold R, Feger KH, Siemer B (16/2006) Standörtliche Potenziale für den Anbau schnellwachsender Baumarten auf Ackerflächen. Allg. Forst. Z. Waldwirtsch. Umweltvorsorge 16/2006:855-857
46. Murach D, Hartmann H, Walotek P (2008) Ertragsmodelle für landwirtschaftliche Dendromasse. Endbericht DENDROM - Zukunftsrohstoff Dendromasse. Berlin, Cottbus: Ebers-walde. pp 146-149
47. Aust C (2012) Abschätzung der nationalen und regionalen Biomassepotentiale von Kurzumtriebsplantagen auf landwirtschaftlichen Flächen in Deutschland, Dissertation an der Fakultät für. Forst- und Umweltwissenschaften: Universität Freiburg.
48. Seitz R (2010) Rinderhaltung in Baden-Wuerttemberg. Statistisches Monatsheft Baden-Wuerttemberg. 30-35
49. Rösch C, Raab K, Stelzer V (2005) Surplus grassland - a new source of bio-energy? In: Proceedings of the 14th European biomass conference & exhibition. Paris: Frankreich. pp 312-315 CD-ROM
50. Landwirtschafts- und Landeskulturgesetz Baden-Württemberg Novellierung vom 17.12.2011. Available at http://www.landesrecht-bw.de/jportal/?quelle=jlink&query=Lw%2FKultG+BW&psml=bsbawueprod.psml&max=true&aiz=true

51. Ministerrat der Europäischen Union (1992) Richtlinie/92/43/EWG vom 21. Mai 1992 zur Erhaltung der natürlichen Lebensräume sowie der wildlebenden Tiere und Pflanzen. Available at http://eur-lex.europa.eu/LexUriServ/LexUriServ.do?uri=CONSLEG:1992L0043:20070101:DE:PDF
52. Oppermann R (2009) Grünland und Schutz von Biodiversität und Gewässern. Vortrag auf der Tagung 'Naturschutz und Landwirtschaft im Dialog - Grünland im Umbruch' am 27.-30.04.2009 an der Internationalen Naturschutzakademie Insel Vilm des Bundesamtes für Naturschutz
53. Richtlinie des Ministeriums Ländlicher Raum zur Förderung der Erhaltung und Pflege der Kulturlandschaft und von Erzeugungspraktiken, die der Marktentlastung dienen (Marktentlastungs- und Kulturlandschaftsausgleich – MEKA II –) vom 12.09.2000. Available at http://www.vd-bw.de/spiderlink/7803-3-1-3alt0708_02.B_inhalt
54. Tonn B, Messner J, Elsässer M (2010) Einsatz von (MEKA-) Grünlandaufwüchsen in Biogasanlagen in Baden-Württemberg, Talk on Expert Workshop. Jena: Einsatz von Grünlandaufwuchs und Landschaftspflegematerial in Biogasanlagen. Available at http://www.tll.de/ainfo/pdf/gland/gn10_06f.pdf
55. Hartmann A (2008) Wie viel Fläche wird für Biogas benötigt? Statistisches Monatsheft Baden-Württemberg 7/2008. 40-42
56. Gömann H, Kreins P, Münch J, Delzeit R (2010) Auswirkungen der Novellierung des Erneuerbare-Energien-Gesetzes auf die Landwirtschaft in Deutschland, talk on the GEWISOLA meeting. Möglichkeiten und Grenzen der wissenschaftlichen Politikanalyse, 29.09-01.10. Braunschweig
57. Breustedt G, Habermann H (2010) Einfluss der Biogaserzeugung auf Landwirtschaftliche Pachtpreise. talk on the GEWISOLA meeting. Möglichkeiten und Grenzen der wissenschaftlichen Politikanalyse. Braunschweig
58. Statistisches Landesamt Baden-Württemberg (2010) Milcherzeugung und Milchverwendung in Baden-Württemberg. Available at http://www.statistik.baden-wuerttemberg.de/veroeffentl/Statistische_Berichte/3432_10001.pdf

Chapter 8

Expected Climate Change Impacts on Land and Natural Resource Use in Namibia: Exploring Economically Efficient Responses

Jon I Barnes[1], James MacGregor[2], and Moira Alberts[3]

[1]Design & Development Services cc, Windhoek, Namibia

[2]Department of Economics, Gothenburg University, Gothenburg, Sweden

[3]Independent consultant, Riebeek-Kasteel, 7307, Western Cape, South Africa

ABSTRACT

This paper informs policy over the efficiency of investments on Namibia's rangelands both now and under future expected climate change. It is in this setting that Namibia's pastoralist communities, communal conservancies and increasing numbers of wildlife reside, and their economic activity dominates. We bring the principles of economics to decision-making. We analyse primary land uses, livestock production and wildlife viewing tourism in three ecologically different biomes - the southern Karoo Biome, the central Savanna Biome and the northeastern Woodland Biome. We analyse how expected climate change will impact these investments to 2080. From an economic perspective, pastoralism and wildlife-based tourism developed within common property management systems can be sound investments across Namibia's communal lands. Furthermore, as climate changes, the existing economic benefits to the Namibian economy associated with pastoralism and wildlife-based tourism should be magnified, as capital-intensive commercial fenced ranching systems become less economic. Recent research indicates that adoption of good rangeland management principles involving flexible herding, economies of scale and commercially viable utilisation of invasive bush can significantly increase profits and economic returns on investment for livestock systems in Namibia. Furthermore, under conditions of expected climate change, our results indicate that income from natural resource use will be more resilient than livestock and other agricultural investments. Estimated losses in economic contribution will be highest in the agricultural sector, and in particular with commercial fenced ranching, and rainfed cropping. Climate change poses considerable challenges for pastoralists and other economic activity on Namibia's rangelands. Global climate modelling suggests that by 2080, Namibia's temperatures will have increased by some 3°C and rainfall decreased by 5% to 20%. Our results suggest that gradual change in climate over the next 70 years will see the potential national value of livestock income halved and that of tourism income reduced by one quarter. Active interventions should include shifts in livestock and rangeland policy, encouraging the adoption of more flexible and resilient systems and including efforts to make rangeland use less rigid and more able to change and adapt. Adaptation to climate change should include promotion of natural resource-based land uses such as continued expansion of the Community Based Natural Resource Management programme. Furthermore, national policy now recognises this, and implementation of the Namibia Rangeland Management Policy will provide incentives to invest in sound rangeland management.

INTRODUCTION

Namibia embraces some 900,000 km^2 and straddles the Tropic of Capricorn on the southwest seaboard of Africa. It is sparsely populated, with one of the driest countries in sub-Saharan Africa, containing biomes ranging from true desert on the west coast, through arid semi-desert and semi-arid savanna, to semi-arid and sub-humid woodlands in the northeast. Land use is based on natural rangelands and takes place commercially on private land, traditionally on communal state land, and to a lesser extent through leasehold in protected areas.

(Reid et al. 2007) used results from a stakeholder workshop exercise and a computable general equilibrium model to estimate likely overall impacts of climate change on the Namibian economy. They reported that over a period of 20 years, the annual losses to the Namibian economy could be up to 6% of gross national product (GDP) due to the impact that climate change will have on its natural resources. They argued that this will affect the poor most, with resulting constraints on employment opportunities and declining wages, especially for unskilled labour, and urged Namibia to take steps that will ensure 'climate proofed' policies and activities and to have a strategy to deal with displaced farmers and farm workers. (Reid et al. 2007) identified the need for climate change to be mainstreamed into policies and planning. (Turpie et al. 2010) have analysed the vulnerability of biodiversity and protected areas in Namibia in the face of climate change. They also examined adaptations expected to result and made recommendations in relation to possible interventions which might mitigate impacts.

(Humavindu et al. 2011) have analysed national-level policy options for the main agricultural and natural resource land uses under the national rangeland management policy (NRMP) (MAWF 2012). This policy is found to be economically efficient since costs associated with implementation of sound rangeland management are recovered through increased financial and economic income. Further, adoption of sound grazing management plans holds potential for significantly increased profits and returns on investment for rangeland users. Furthermore, as time goes on, with increased adoption and improved skills and understanding among range users, national income amounting to some N$5 billion (US $610 million at September 2012 prices) might be added to the national economy from the rangeland use sector. Such benefits can go a long way towards mitigating and reversing the anticipated negative effects of climate change and increasing drought incidence in Namibia.

The present study, conducted in parallel with that of (Turpie et al. 2010), reports on a specific study on the impacts of climate change on the main agricultural and natural resource land uses of Namibia (the study area). It examines these land uses from a financial and economic point of view and assesses the changes in direct economic contribution that could be expected to result from climate change. The likely impacts of expected changes in climate were interpreted in terms of their effect on the natural resource base. The resulting changes were applied using sensitivity analysis in land and natural resource use enterprise models to determine effects on the profitability and economic contribution of these activities. Most analytical emphasis was placed on primary land uses, livestock production and wildlife viewing tourism in three ecologically different biomes - the southern Karoo Biome, the central Savanna Biome, and the northeastern Woodland Biome. However, all land and natural resource uses in these areas were considered. These are described below.

Crop and Livestock Production

Livestock are the mainstay of agricultural production and land use in Namibia. The Woodland and Savanna biomes, found in the north east, north and centre, are essentially dominated by cattle, with lesser amounts of small stock. The Karoo biome found in most of southern Namibia is dominated by small stock (sheep and goat) production.

All livestock production is based on extensive systems, breeding and rearing or rearing of stock, based on the productivity of natural rangeland. Cattle are kept on privately owned or leased commercial land on medium-scale fenced farms averaging some 15,000 ha each to produce beef. They are also kept on state-owned communal land, mostly in unfenced common property systems on a small-scale around villages to produce meat, milk, as transport and as a store of wealth. They are also kept at medium-scale cattle posts away from villages for the production of beef and as a store of wealth. Some commercial production in fenced land is taking place in communal land mainly to produce beef and as a store of wealth. Fenced commercial farms in the arid south are used to produce mutton sheep and karakul pelts with small amounts of wool, and goats are kept for meat production. On communal land in the south, small-scale production of sheep and goats, also for mutton, pelts and goat meat, takes place on open access or partly fenced properties. Traditional nomadic and semi-nomadic livestock keeping in the arid west and south of the country is no longer prevalent, and almost all extensive livestock keeping on rangeland is sedentary around fixed water points. Relatively unimportant livestock systems include intensive dairy, pig and ostrich production on commercial land.

Livestock stocking rates on rangeland are maintained by commercial farmers at around the most biologically productive carrying capacities (between 10 and 15 ha per large stock equivalent unit in the north and northeast, and between 15 and 25 ha per large stock equivalent unit in the south and extreme west). In communal common property systems, the stocking rates are closer to the biological limits, the ecological carrying capacity (between 5 and 7 ha per large stock equivalent unit in the north and northeast, and between 7 and 12 ha per large stock equivalent unit in the south and extreme west). Herd and flock productivities (calving/lambing rates, mortality rates and animal growth rates) tend to reflect the stocking rates, highest in commercial land and lowest in the communal open access systems. Cattle post systems tend be between the two in terms of stocking rates and productivity. The carrying capacities referred to here are derived from (Mendelsohn et al. 2006), and reflect empirically established norms of the Ministry of Agriculture Water and Forestry.

Both dryland (rainfed) and irrigated crop production are practised in Namibia. Dryland crops are grown in association with small-scale livestock production within agro-pastoral systems in the north and northeast. Here, low-input, small-scale production of millet, maize and sorghum, sometimes with associated field beans, are grown. Yields are low, and on average, in one of three years, crops fail. Households producing crops tend to produce less than half of their annual subsistence needs, relying on cash purchases to fill the gap. Medium-to large-scale dryland crop production takes place on commercial land, mostly in the northeast, where maize is the primary crop.

Irrigated crop production takes place on commercial land in the south along the Orange and Fish rivers, in communal land in the northeast along the Okavango river, and to a lesser extent, in the northwest, associated with the Kunene river. A variety of crops is produced, including a mix of food crops and higher value cash crops. In the south, table grape production is very significant. Also grown there are fodder crops and maize, among numerous others. In the north and east, in mostly state-run schemes, the emphasis has been on food crops, but increasing amounts of higher value crops are being introduced. The Green Scheme is an ambitious state-driven initiative aimed at significantly boosting irrigated crop production in the country. A notable feature of irrigated agriculture is that schemes in the south tend to be financially and economically viable, while those in the north tend to have poor or non-existent financial economic viability. This is largely due to the location, remote from markets for inputs and products of the northern schemes (Barnes et al., unpublished work).

Tourism, Wildlife Production and Natural Resource Uses

Tourism is a rapidly growing sector in Namibia, and the leisure tourism component of this, which makes up some 40%, is dominated by nature-based pursuits. The nature-based component is attributable to a range of natural assets, including scenery and wildlife, which are the most important. (Barnes et al. 2009) used data from the tourism survey of (SIAPAC 2007) to estimate that wildlife contributed some 50% of the value of nature-based tourism. Approximately 80% of the value of nature-based tourism is non-consumptive, based in lodge or camp accommodation in the commercial land, state land (parks), and communal land. The other 20% consists of consumptive tourism, dominated by trophy hunting, which takes place mostly on commercial land but also on communal land and state land. Another, lesser, consumptive tourism component consists of marine shore angling.

Tourism on commercial land takes the form of landholder enterprises, while that on state and communal takes the form of leasehold enterprises in joint venture partnerships with land holders - either government

or local communities. The well-established community-based natural resource management (CBNRM) programme enables communities on communal land to form common property conservancies through which they can manage and utilise wildlife mostly for tourism. Tourism as described above holds significant comparative advantages for Namibia as it is not entirely dependent on scarce and erratic rainfall, but also makes use of the natural beauty inherent in the landscapes. Tourism potential tends to be localised around areas of high scenic value and high wildlife concentrations.

The use of wildlife other than through tourism takes place on commercial land through consumptive harvesting of game for own use, small-scale meat sale, live game capture and sale, and medium-scale commercial meat sale. Most of this use takes place as supplementary enterprises alongside livestock production or tourism enterprises. Wildlife numbers have been increasing relative to livestock numbers, which have generally been declining on commercial land. This is a function of higher perceived potential returns, which have induced investments in wildlife within an improved property rights setting (Barnes and Jones 2009). Wildlife use on commercial land ends to be fenced at least at large scale, while that on communal land is generally unfenced. Although wildlife-based and livestock land uses commonly take place together at regional and landscape scale, they tend to be separated spatially at farm/conservancy and enterprise scale.

The use of wild plant resources takes place mostly on communal land and to a lesser extent on commercial land. Nearly all rural households harvest wood for fuel, poles for building, and non-timber wild plant products for food, medicines and as raw materials for crafts. A significant charcoal industry is developing on commercial land where bush encroachment has been a problem. Inland fishing is practised in the wetlands of the northeast and north. Here, households operate at a small scale to harvest fish from rivers and floodplains. The use of natural resources on communal land is generally undertaken by individual households in an open access setting. In a small number of areas, common property management is being applied through community-based organisations such as conservancies and community forests.

METHODOLOGY

Review of Climate Change Predictions

The evidence for anthropogenic climate change as it applies to Namibia was reviewed. Much of the work already done in this region is based on older generation climate scenarios dating from the early 2000s, and the Third Assessment Report of the Intergovernmental Panel on Climate Change (IPCC TAR). Later generation climate change scenarios have been developed for the Fourth Assessment Report (IPCC AR4). Key references were GF Midgley and D Gou (unpublished work), (Turpie et al. 2010), (Midgley et al. 2005), who used these scenarios in the context of Namibia.

Development and Analysis of Land Use Enterprise Models

The livestock systems and tourism wildlife viewing enterprises in Namibia were modelled to provide financial and economic measures of value. Standardised financial and economic, spreadsheet, enterprise appraisal models as developed in the economics unit of Ministry of Environment and Tourism (MET) were developed for each enterprise. These include annualised budgets and 10-year enterprise development streams valued in Namibia dollars (N$) at constant 2008 prices, where N$1:00 = US$0.13.

The financial analyses measured the *profitability* for investors or the financial incentive to engage in the activity. The economic analyses measured the value added by these activities to the national income, measured as the *direct* net change in *gross national income*. No account was taken of the indirect value added associated with the activities, though backward and forward linkages. Nor was account taken of indirect use values (ecosystem service values), and non-use values (existence, bequest and option values).

For the purpose of this study, the parameters, assumptions and estimations for the models were derived from work done in Botswana and Namibia at various times. Parameters for the Savanna and Woodland biome livestock models were determined using data from (Barnes et al. 2008) and Barnes et al., unpublished work). The parameters for the wildlife viewing tourism models in the Savanna biome were determined using data from (Massyn et al. 2008). The Woodland biome tourism parameters were derived from (Alberts 2007), and the Karoo biome livestock and tourism parameters were derived from JI Barnes and MN Humavindu (unpublished work).

These were then valued to estimate financial values for initial and ongoing capital requirements, recurrent variable and fixed operating expenses, and recurrent enterprise gross income. The annual net income (profit), income accruing to local communities, was determined. The overall financial return on investment was measured as the 10-year internal rate of return and the 10-year net present value. Tables 1, 2, 3 and 4 show some of the parameters used in the enterprise models.

Table 1: Comparative key parameters used in base case models for the Woodland and Savanna biome livestock systems

Parameters	Traditional small scale	Cattle post	Commercial ranch
Land used (ha)	180	6,400	10,000
Number of cattle (head)	35	760	1050
Number of goats (head)	3	59	100
'Economic' carrying capacity (ha/LSU)	13	13	13
Stocking rate (ha/LSU)	6.5	8.3	12.9
Calving rate (% cows)	60	63	65
Calving rate (% heifers)	60	63	65
Mortality rate (% calves)	18	9	5
Mortality rate (% others)	11	5	3
Bull rate (% herd)	5	5	5
Goat reproductive rate (%)	20	25	30
Average total herd growth rate (%)	0.7	8	4.6
Cattle off-take rate (% herd)	9	12	18
Goat off-take rate (% flock)	20	25	30

Milk yield (litres/lactating cow/annum)	158	n/a	n/a
Transport/draft (days/span of four/annum)	55	n/a	n/a

LSU = Large stock unit.

Barnes *et al.*

Barnes *et al. Pastoralism: Research, Policy and Practice* 2012 2:22, doi:10.1186/2041-7136-2-22

Table 2: Comparative key parameters used in base case models for the Karoo biome livestock systems

Parameters	Traditional small scale	Commercial ranch
Land available (ha)	87,000	87,000
Land used (ha per enterprise)	2,330	52,200
No. livestock (head)	285	10,346
No. large stock units (LSU)	47	871
Stocking rate (ha/LSU)	50	60
'Economic' carrying capacity (ha/LSU)	60	60
Lambing/kidding rate (% of ewes)	90	130
Lambing/kidding rate (% of young ewes)	98	130
Mortality rate (% of lambs)	40	28
Mortality rate (% of adults)	7	5
Ram rate (% of flock)	3	3
Stock off-take rate (% of stock numbers)	16	31
Biomass production (% of stock LSU)	14	24

Barnes *et al.*

Barnes *et al. Pastoralism: Research, Policy and Practice* 2012 2:22, doi:10.1186/2041-7136-2-22

Table 3: Comparative key parameters used in base case models for tourism wildlife viewing enterprises in the Woodland, Savanna and Karoo biomes (2008)

Parameters	Woodland	Savanna	Karoo
Land used (ha)	14,400	27,000	87,000
Game density	6.23	3.11	0.96
Core wildlife area	n/a	16,835	54,245
'Economic' carrying capacity (ha/LSU)	13	13	60
Hectares per tourist bed	800	218	613
Number of lodges	1	3	4
Number of tourist beds	18	124	142
Occupancy rate (per annum)	0.54%	0.63%	63%
Average length of stay	3	4	4
Daily tariffs - up-market lodge (N$)	3,580	3,000	647
Daily tariffs - roadside lodge (N$)	n/a	1,300	452
Daily tariffs - mid-market lodge (N$)	n/a	1,300	300
Daily tariffs - camping (N$)	n/a	n/a	172

Barnes *et al.*

Barnes *et al. Pastoralism: Research, Policy and Practice* 2012 2:22, doi:10.1186/2041-7136-2-22

Table 4: The estimated extent of land divided according to tenure in Namibia

Extent km2	Total
State land	12,360,000
Communal land	34,608,000
Commercial land	35,432,000
Total extent	82,400,000

Barnes *et al.*

Barnes *et al. Pastoralism: Research, Policy and Practice* 2012 2:22, doi:10.1186/2041-7136-2-22

The livestock systems modelled for the Woodland and Savanna biomes included unfenced traditional small-scale livestock keeping on open access communal land, unfenced cattle posts, medium-scale ranching on partially exclusive communal land and fenced commercial medium- to large-scale ranching (breeding and rearing) on fully exclusive commercial land. All these were dominated by cattle and entailed some form of breeding and rearing. The traditional small-scale system is typical of the more densely settled areas of communal land where stocking rates are high and overall herd growth is slow, while the medium-scale cattle post system is expanding into open communal land where stocking rates are moderate and the overall herd growth is high. In the fenced commercial systems, stocking rates are maintained at near optimal levels for biological productivity, allowing for highest herd productivity and moderate herd growth.

In the Karoo biome, the livestock systems, dominated by small stock, included unfenced traditional small-scale small stock production on communal land and fenced large-scale commercial small stock ranching on commercial land. These involved breeding of karakul sheep for pelts, and or breeding and rearing of sheep and goats for meat production.

The tourism systems modelled in all biomes involved exclusive unfenced commercial large-scale non-consumptive tourism with lodge development on communal land. The models involve a leasehold system where the tourism investors do not own the land or wildlife stock. Generally, lower livestock stocking rates, on commercial land and in the open parts of communal land, enable wildlife populations to survive and provide the basis for investment in wildlife conservation and use through tourism. As stated, financial and economic comparative advantage has ensured a steady relative increase in wildlife-based land use in these areas (Barnes and Jones 2009).

The assumptions used in the livestock system models differed slightly between those for the Woodland and Savanna biomes. However, the differences were small, as the higher primary productivity in the wetter northeastern woodlands tends to be offset by less favourable vegetation structure for livestock grazing.

The use of enterprise models for sensitivity analysis was restricted to the livestock and tourism land uses described above. For other land and natural resource uses, enterprise models were used subjectively to assess the likely impacts of climate change on their net incomes and economic contributions. Models for several land and natural resource uses were derived from Barnes et al. (unpublished work). These included models for small- and medium-scale rainfed crop production, as it takes place in the north and northeast of the country, large-scale commercial irrigation along the perennial rivers and small-scale inland fishing enterprises. Models for wild plant use were derived from (Barnes et al. 2005, 2010). The models developed of community-based natural resource management (CBNRM) initiatives by (Barnes et al. 2002) were also used to examine the resilience of community conservancies to changes in rangeland productivity.

Aggregate Economic Value of Agricultural Production and Natural Resource Use

The land uses described above have been valued in terms of their contribution to the gross national income. This is the baseline for measuring the likely economic impacts of climate change. The study focused on the land use activities. Only their direct impacts on the national economy were measured, and no account was taken of backward and forward linkages associated with these land and natural resource uses. Thus, value added through enterprises, supplying inputs to the land and resource use activities, or value added as a result of product processing such as meat processing, is excluded. The values are given in value added to the gross national income (GNI).

The basis for the agricultural values is the national accounts' GDP by activity data, presented by IPPR [Institute for Public Policy Research] (2009), which cover the period between 1990 and 2005, divided into commercial and communal agriculture. With the use of ratios from (Sherbourne 2009) and (Mendelsohn et al. 2006), the data were disaggregated into those attributable to commercial land livestock, communal land livestock, commercial land crop production and communal land crop production. The commercial land livestock value was further disaggregated into cattle and small stock. These values were projected to 2009 using the averages of real annual growth rates recorded between 1995 and 2005 and then inflated as appropriate to reflect 2009 prices. The annual growth rate applied for the commercial sector was 3.4% and that for communal sector was 2.1%.

Natural resource values were derived from various sources. Values for nature-based tourism, including trophy hunting tourism, were derived from the wildlife accounts (Barnes et al. 2009) and verified through comparison with the studies by (Turpie et al. 2010) in the value of parks, and the tourism satellite accounts WTTC/World Travel and Tourism Council (2006) and (Barnes et al. 2010, 2008). These values were subjected to real growth of 5.5% per annum to 2009, reflecting growth rates recorded and predicted by (WTTC 2006) and adjusted to take account of the 2008 economic downturn. They were then also inflated to reflect 2009 prices. Values for wildlife use other than tourism were derived from the wildlife accounts (Barnes et al. 2009). They were adjusted to 2009 values, through real growth at 7.2% per annum, based on average growth between 2004 and 2007, recorded by T Uahengo, (personal communication), adjusted to take account of the 2008 economic downturn and thereafter inflated accordingly.

Values for the use of wild plants were derived from the forest accounts which include the use of non-timber wild plant products (Barnes et al. 2005). They were adjusted to 2009 figures using the expected real annual growth rate of 4.2% predicted in the accounts, adjusted to take into account the 2008 economic downturn and inflated accordingly. Values for inland fish use were derived from the 2008 values estimated by Barnes et al. (unpublished work), extrapolated to all areas with inland fisheries and inflated accordingly. These values then disaggregated according to those attributable to state land (protected areas), communal land and commercial land. This was done using areas estimated using a dot counting technique and modified subjectively according to the knowledge regarding the distribution of resources and land uses.

Aggregate values were thus described according to the three integrated regions subdivided between state, communal and commercial lands. The extents of these different tenure types in the country were estimated, as shown in Table 1, and values were then expressed per square kilometre within each land tenure category, again divided between the three integrated regions.

Next, using the results from the sensitivity analysis of land use enterprise models, the impact of likely climate change on carrying capacities for livestock and wildlife and on tourist visitor numbers, as described in the analysis above, was assessed. Again, the measure of economic value is the direct contribution of land and resources to the GNI. The impact is assessed in terms of how much of the baseline direct GNI values, as measured above, will change by 2080 as a result of the climate change predictions. It is noteworthy that this analysis was done by superimposing the 2080 setting after climate change on current land use. It does not take into consideration autonomous adaptations that are expected to happen, or adaptations resulting from policy interventions, or changes in population growth, which will happen between the present and 2080.

The challenge was to convert the predicted changes in physical livestock, wildlife and tourist numbers, referred to above, into changes in GNI contribution. For this, we used detailed spreadsheet enterprise models for livestock, tourism and natural resource use activities. These empirically based budget and cost-benefit models measure, among other things, the annual direct contribution to GNI. With the use of sensitivity analysis, the impacts of changes on carrying capacity, tourism numbers and output on GNI contribution can be measured. Specifically, the models on land use developed in this study were used. The latter study used enterprise models for 11 different land uses, examining communal and commercial land livestock systems and non-consumptive tourism systems.

RESULTS

Climate Change Predictions

GF Midgley and D Gou (unpublished work) used model results from IPCC AR4 (the United Nations Intergovernmental Panel on Climate Change Fourth Assessment Report) to estimate the likelihood of the outcomes estimated using older scenarios. They noted that the model projections of the earlier IPCC TAR (the United Nations Intergovernmental Panel on Climate Change Third Assessment Report), used by (Midgley et al. 2005), are broadly comparable with median projections assessed in AR4, in that a similar level and spatial pattern of warming is consistent and that a comparable reduction in rainfall is projected. In the present study, we have used the findings of GF Midgley and D Gou (unpublished work) as the basis for our model projections.

The models used (IPCC TAR and AR4) are extremely complex global models which require very large numbers of runs to develop predictions. At the scale of southern Africa, they are thus coarse, and it is difficult to get the spatial downscaling needed to make accurate predictions at country level in Namibia. Nevertheless, averages of 21 IPCC AR4 models show a consistent trend to higher temperatures, with an increase in annual mean temperature between 3°C and 4°C expected by 2080. Regarding precipitation, the picture is less certain with the averages for 21 IPCC AR4 models showing decreases in mean annual rainfall between 5% and 20% by 2080. The most significant declines can be expected in the south and centre of the country. This is consistent with the continent-wide finding that it is the southwestern arid zone of southern Africa that will be likely to suffer high declines in rainfall compared to the wetter parts, to the east of Namibia. However, the dominant feature of the predicted changes, particularly as regards precipitation, is an increase in variability.

The effects of higher temperature will result in higher evaporation, and this, combined with expected lower rainfall, would cause significant reductions in rangeland carrying capacity. In the analysis on impacts on ecosystems and biodiversity, GF Midgley and D Gou (unpublished work) suggest a reduction in ground cover and reduced net primary productivity throughout much of the country by 2080. They estimated changes in grass biomass ranging from around −10% in the Woodland of the northeast, around −5% in the Savanna and around −10% in the southern Karoo areas. They estimated that primary productivity overall might have no change in the Woodland and might drop by some 5% in the Savanna and Karoo biomes. These indicated changes, by 2080, have been used in the interpretation of carrying capacity changes below.

It must be acknowledged that increased concentrations of CO_2 may result in a fertilisation effect as predicted by the dynamic global vegetation model described by (Scheiter and Higgins 2009). This modelling approach uses climate and soils information to predict the changing relative success of a few main plant functional types. The model has a strong physiological basis, including calculations of carbon uptake via photosynthesis (dependent on light, water, temperature and nutrient availability), respiration and growth, carbon allocation to above and below ground parts, decomposition and wildfire. The model has been shown to simulate African savanna and grassland vegetation structure and function to a credible degree (Scheiter and Higgins 2009). As with all global climate change models, there are difficulties with the spatial downscaling to make clear predictions for Namibia, but the results indicate small expansion of the grass biomass conditions prevailing in the south into the central areas of the country by 2080. In the north central and northeast, they indicate possible increases in grass biomass. They also suggest a decrease in woody plants less than 2 m tall in most of the central and northern parts. For woody plants more than 2 m tall, little change or possibly an increase is suggested for the central and northern areas.

It is uncertain whether CO_2 fertilisation will increase plant biomass. This effect, if any, could be expected in the wooded north and northeast, but it would be offset by the ultimate changes in precipitation, if any. As stated, for the interpretation of carrying capacity changes, below, we used the changes indicated by (Midgley et al. 2005). It has been estimated by (Turpie et al. 2010) that these changes would likely be accompanied by losses in both grazer and browser carrying capacities of between 13% and 15%.

Surface water flows are predicted to decline by up to 15% in the Orange river system to the south and 10% in western ephemeral rivers. There is less certainty about flows in the north and northeast flowing out of Angola, but some predictions suggest decline, which will also affect the wetlands associated with these systems. Combined with significantly increasing demand, the pressure on Namibia's surface water resources will be dramatic.

Financial and Economic Analysis of Land Use Enterprises

Tables 5, 6 and 7 show the base case financial and economic characteristics of the livestock and tourism enterprises modelled. Generally, traditional small-scale cattle keeping, as practised in the Woodland and Savanna biomes, is shown to make efficient use of capital and to provide significant income for participating households. However, these returns tend to be dependent on the maintenance of reasonable levels of herd productivity. Open access to grazing, which commonly prevails with this system, has the effect of driving down herd productivity and the net economic contribution (Barnes et al. 2008). The diversity of products lends stability to the system. In the Karoo biome, traditional small-scale small stock keeping, represents a small commercial operation. Here, diversity of products is low, and the system suffers from diseconomies of scale coupled with typically low incomes earned from open access grazing. Improvement of flock productivity while keeping capital inputs low would considerably enhance the value of this system.

Table 5: Financial and economic values associated with livestock systems and tourism wildlife viewing enterprises – Woodland biome (N$, 2008, where N$1.00 = US$0.13)

	Traditional	Cattle post	Commercial	Tourism
Financial data				
Initial capital	99,381	1,064,782	3,259,660	11,825,946
At stability				
Gross turnover	35,225	380,143	937,506	15,076,056
Financial costs variable	8,050	241,301	418,605	7,701,617
Financial costs fixed	10,677	81,713	424,152	3,089,850
Entrepreneur net cash income	16,498	57,129	94,749	4,284,590
Community income	23,518	27,000	62,370	626,400
Overall return				
Financial IRR (10years)	19.99%	7.79%	7.09%	35.05%
Financial NPV (10years at 10%)	48,792	(19,245)	(570,856)	17,509,309
Wage bill	7,020	27,000	62,370	1,144,800
Economic data				
Initial capital	86,756	958,603	2,979,933	10,814,706
Gross output	29,372	432,123	814,223	16,335,133
Economic costs	19,549	263,867	608,497	9,368,723

Incremental contribution to GNI	9,823	168,256	205,726	6,966,410
Incremental contribution to NNI	7,963	145,309	69,992	6,292,706
Economic IRR (10 years)	23.94%	11.53%	7.49%	70.29%
Economic NPV (10 years) at 10%)	56,184	99,006	(409,815)	34,391,437
Number of jobs	1.3	4	5	28
Incremental contribution to GNI per hectare	54.57	26.29	20.57	483.78
Incremental contribution to NNI per hectare	44.24	22.70	7.00	436.99
Community income per hectare	130.66	4.22	6.24	43.50
Entrepreneur net cash income per hectare	91.66	8.93	9.47	297.54
Initial economic capital per hectare	481.98	149.78	297.99	751.02
Economic gross output per hectare	163.18	67.52	81.42	1,134.38

IRR, internal rate of return; GNI, gross national income; NNI, net national income; NPV, net present value.

Barnes *et al.*

Barnes *et al.* *Pastoralism: Research, Policy and Practice* 2012 2:22, doi:10.1186/2041-7136-2-22

Table 6: Financial and economic values associated with proposed livestock systems and tourism wildlife enterprises - Savanna biome (N$, 2008, where N$1.00 = US$0.13)

	Traditional	Cattle post	Commercial	Tourism
Financial data				
Initial capital	87,345	1,019,233	3,316,252	34,902,460
At stability				
Gross turnover	29,237	374,960	828,721	46,051,817
Variable financial costs	7,802	223,486	361,592	35,102,546
Fixed financial costs	10,017	80,864	425,608	7,434,437
Entrepreneur net income	10017	70,610	41,521	3,514,834
Community income	18,438	97,610	62,370	1,487,160
Overall return				
Financial IRR (10years)	15.49%	8.30%	6.59%	21.02%
Financial NPV (10years at 10%)	24,733	(135,447)	(637,192)	15,202,591
Wage bill	7,020	27,000	103,950	3,287,160
Economic data				
Initial capital	75,501	894,863	3,068,388	30,307,333
Gross output	24,893	426,422	673,377	41,877,101
Economic costs	19,136	256,154	584,119	31,476,663
Incremental contribution to GNI	5,757	170,268	89,258	10,400,439
Incremental contribution to NNI	4,536	147,025	(47,474)	9,577,954
Economic IRR (10years)	19.76%	10.86%	6.81%	63.09%
Economic NPV (10years at 10%)	35,386	56,443	(501,139)	55,517,848
Number of jobs	1.3	4	7	135

Incremental contribution to GNI per hectare	31.99	24.05	8.93	385.20
Incremental contribution to NNI per hectare	25.20	20.42	(4.75)	354.74
Community income per hectare	102.43	8.12	6.24	55.08
Entrepreneur net cash income per hectare	63.43	7.03	4.15	130.18
Initial economic capital per hectare	419.45	140.25	306.84	1,122.49
Economic gross output per hectare	138.30	66.63	67.34	1,551.00

IRR, internal rate of return; GNI, gross national income; NNI, net national income; NPV, net present value.

Barnes *et al.*

Barnes *et al. Pastoralism: Research, Policy and Practice* 2012 2:22, doi:10.1186/2041-7136-2-22

Table 7: Financial and economic values associated with proposed livestock systems and tourism wildlife viewing enterprises – Karoo biome (N$, 2008, where N$1.00 = US$0.13)

	Traditional	Commercial	Tourism
Financial data			
Initial capital	139,384	7,481,255	6,527,865
At stability			
Gross turnover	25,208	1,972,255	6,627,604
Financial costs variable	13,775	612,625	12,130,743
Financial costs fixed	8,095	1,102,375	9,056,362
Entrepreneur net income	3,338	257,255	5,440,499
Community income	6,578	219,024	2,577,744
Overall return			

Financial IRR (10 years)	5.42%	13.42%	19.81%
Financial NPV (10 years at 10%)	(29,844)	1,129,712	17,984,661
Wage bill	3,240	453,024	3,981,744
Economic data			
Initial capital	119,933	7,386,407	30,350,760
Gross output	29,647	2,144,615	25,504,896
Economic costs	23,149	1,650,267	15,994,383
Incremental contribution to GNI	6,498	494,348	9,510,514
Incremental contribution to NNI	3,857	308,247	8,268,390
Economic IRR (10 years)	18.48%	27.24%	40.03%
Economic NPV (10 years at 10%)	42,544	4,190,185	44,289,321
Number of jobs	0.60	29	153
Incremental contribution to GNI per hectare	2.79	9.47	109.32
Incremental contribution to NNI per hectare	1.66	5.91	95.04
Community income per hectare	2.82	4.20	29.63
Entrepreneur net cash income per hectare	1.43	4.93	62.53
Initial economic capital per hectare	51.47	141.50	348.86
Economic gross output per hectare	12.72	41.08	293.16

IRR, internal rate of return; GNI, gross national income; NNI, net national income; NPV, net present value.

Barnes *et al.*

Barnes *et al. Pastoralism: Research, Policy and Practice* 2012 2:22, doi:10.1186/2041-7136-2-22

Commercial cattle ranching, mainly in the commercial land of the Woodland and Savanna biomes, and commercial small stock ranching in the Karoo biome all represent relatively capital intensive systems, with fencing and water provision being important assets. They depend on gains in productivity resulting from relatively intensive infrastructure and management inputs, and their net returns are sensitive to cost increases. Cattle post systems are less intensive with lower capital requirements and less dependence on proximity to markets, and they benefit from only moderate gains in productivity that resulted from their partial exclusivity.

As described above, the likely general effects of climate change in Namibia will be higher temperatures and likely lower mean annual rainfall. These will combine to increase evapotranspiration and reduce rainfall effectiveness. This should reduce primary productivity, but this reduction *may* be offset to some extent by CO_2 fertilisation. To go with these changes, there is strong evidence that there will be greater variability in climate and thus in ecological systems, which are already highly variable.

The land uses described above will be affected by reductions in rangeland carrying capacities. Before systems adapt, there will likely be effects of livestock production parameters such as reduced calving rates and higher mortalities. The sensitivities for these parameters are thus highly significant. Lower production in the face of increasing demand for products may tend to increase meat prices, particularly locally in Namibia. The lower production per unit of land will also tend to increase capital costs relative to production capacity. It is noteworthy that these initial analyses disregard the fact that, by 2080, some 70 years into the future, there may be gradual adaptation to change, through the emergence of more extensive systems and lower production per hectare. Such adaptation is discussed below.

A study of the likely impact of climate change on the gross incomes of landholders practising livestock production and trophy hunting in the commercial land was carried out by (Brown 2009). She used regression analysis on questionnaire returns for a sample of 60 commercial farmers to get farmer perceptions on how their incomes would be affected in the face of climate change scenarios, similar to those identified in this study. For fenced commercial livestock production, it was estimated that gross incomes would decline to some 40% of current levels by 2080, and trophy hunting gross incomes would decline by more. Trophy hunting is more profitable, as an enterprise, than livestock ranching, and a drop in gross income means less in terms of a drop in net returns for trophy hunting than is the case with livestock. For the purposes of the sensitivity analyses on livestock systems, changes in gross output along these lines were examined.

As little is known of the likely impacts of climate change on tourism, the results of a specific survey among tourists using protected areas, described by (Turpie et al. 2010), were used to predict the expected impacts of climate change on tourism visitation numbers. A demand function, derived by (Turpie et al. 2010) from their protected area tourism survey data, indicated that, nationwide, the numbers of tourist visits will be likely to drop to some 85% of the current numbers by 2080, as a result of climate change. A 15% loss in gross output was examined in the sensitivity analyses for tourism enterprises.

Some interesting findings emerge from the sensitivity analyses. The possible effect of increased meat product prices will tend to be offset by increased capital costs in all the livestock systems analysed. In all biomes, losses of 40% of gross incomes will make fenced commercial ranching nonviable both financially and economically. With cattle posts in the Woodland and Savanna biomes, the effect will not be as severe, but financial and economical viability will be reduced significantly. In the case of traditional livestock keeping, the effects of climate change on financial and economic viability will be least. This underlines the inherent efficiency of these systems, which is only negated by the problem of open access grazing, which, as expected from theory and as shown by (Barnes et al. 2008), can drive down productivity. The viability of traditional small-scale livestock keeping in the Karoo biome is, on the other hand, relatively sensitive to climate change effects, reflecting the nature of this system, which tends to resemble small-scale commercial ranching.

In the tourism systems analysed, relatively small losses in financial and economic viability will result from climate change. This makes sense on several grounds. The tourism product is only partially made up by wildlife, with some 50% of it related to scenery and other nonbiological attributes. This is particularly the case not only in the arid Karoo biome, but also in the Woodland biome. Perhaps related to this, the (Turpie et al. 2010) survey of tourists estimated that tourist numbers (and gross outputs) might drop by only 15%. The basic financial and economic viability of the tourism enterprises tends to be higher than

it is for the commercial livestock systems, minimising the effects of income losses.

Baseline aggregate economic values of land and natural resource use

The estimated direct economic contributions in terms of value added to the national income by agricultural land use and natural resource uses in Namibia are documented in Table 8. Table 9 shows these values expressed per square kilometre.

Table 8: Estimated direct gross national income contributed by agricultural production and natural resource uses in 2009

Sector and tenure	GNI contribution (N$'000, where N$1.00 = US$13)
Agriculture	
Livestock	
State land	-
Communal land	521,000
Commercial land	1,965,000
Total	2,486,000
Crops	
State land	-
Communal land	317,000
Commercial land	431,000
Total	748,000
Total agriculture	
State land	-
Communal land	837,000
Commercial land	2,397,000
Total agriculture	3,234,000
Natural resources	
Tourism	
State land	433,000

Communal land	101,000
Commercial land	1,266,000
Total	1,800,000
Wildlife (other)	
State land	72,000
Communal land	15,000
Commercial land	140,000
Total	226,000
Wild plants	
State land	-
Communal land	1,334,000
Commercial land	352,000
Total	1,686,000
Fish (inland)	
State land	-
Communal land	95,000
Commercial land	-
Total	95,000
Total natural resources	
State land	505,000
Communal land	1,545,000
Commercial land	1,758,000
Total natural resources	3,807,000

Barnes et al.

Barnes et al. Pastoralism: Research, Policy and Practice 2012 2:22, doi:10.1186/2041-7136-2-22

Table 9: Estimated direct gross national income contributed per square kilometre by agricultural production and natural resource uses in 2009

Sector and tenure	GNI contribution (N$/km2, where N$1.00 = US$0.13)
Agriculture	
Livestock	

State land	-
Communal land	15.04
Commercial land	55.47
Total	30.17
Crops	
State land	-
Communal land	9.16
Commercial land	12.18
Total	9.08
Total agriculture	
State land	-
Communal land	24.20
Commercial land	67.64
Total agriculture	39.25
Natural resources	
Tourism	
State land	35.06
Communal land	2.92
Commercial land	35.72
Total	21.85
Wildlife (other)	
State land	5.79
Communal land	0.44
Commercial land	3.95
Total	2.75
Wild plants	
State land	-
Communal land	38.54
Commercial land	9.94
Total	20.46
Fish (inland)	
State land	-
Communal land	2.75
Commercial land	-
Total	1.15

Total natural resources	
State land	40.86
Communal land	44.64
Commercial land	49.61
Total natural resources	46.21

Barnes *et al.*

Barnes *et al. Pastoralism: Research, Policy and Practice* 2012 2:22, doi:10.1186/2041-7136-2-22

In terms of tourism, wildlife and natural resource uses, some N$3.8 billion (US$ 490 million) was generated in 2009. This is dominated by two main components - tourism (47%) and natural plant use (44%). Currently, 70% of tourism value is generated in the commercial land, where property rights have allowed a rapid expansion of nature- and wildlife-based land uses. There remains significant potential for tourism development within parks and communal areas. Some 79% of natural plant use takes place in the communal areas, mostly in the north central part of the country (North West broad integrated region) where almost all rural households make use of wood and other natural products of plant origin. There is only localised overuse of these resources, and overall, they are relatively abundant, making for high value added and rent generation, despite the fact that use of the resources is effectively open access in nature.

It can be seen that agricultural production land uses generated some N$3.23 billion (US$ 420 million) in terms of gross national income value added to national income. Some 77% of this is attributable to livestock. Commercial land contributes some 74% to total agricultural land use production value and 79% of livestock production value.

Expected Changes in Aggregate Economic Value

The findings above combined with the findings of (Turpie et al. 2010) were used to predict losses in livestock and wildlife numbers in Namibia, resulting from climate change. Reduced carrying capacity and shifts in land suitability mean that large stock numbers will be particularly hard hit by climate change. The analysis indicates that,

nationwide, numbers of cattle in fenced ranches can be expected to decrease to as low as 40% of the current levels by 2080. Numbers of communal livestock (small-scale and cattle post systems) can be expected to decrease to as low as 70% of the current numbers by 2080. Small stock, mostly in commercial areas, will tend to expand to the north, and numbers of small stock can be expected to remain around 100% of the present numbers.

Numbers of grazing wildlife species (the bulk of current wildlife biomass, particularly in the commercial areas) are expected to decrease to about 75% of the current levels. Disregarding expected growth in wildlife numbers in underutilised areas, the gross output for other (non-tourism) wildlife use can also be expected to drop to 75% of the current levels by 2080. Wild plant use, although highly valuable and suffering locally from depletion, is generally well under the national potential for use of these resources (Barnes et al. 2005). The uses are mainly carried out for essential communal land household livelihood strategies. Disregarding the fact that these uses are likely to increase as other options through agriculture are diminished, it is assumed for this impact analysis that the gross outputs for these activities will remain at 100% of the current levels by 2080. Similarly, inland fish production is treated as remaining at 100% of the current levels by 2080. Nationwide losses in numbers of livestock and wildlife, combined with the losses in financial and economic viability detected at enterprise level, provided the basis for the estimates below of losses in aggregate contribution to GNI by livestock and wildlife-based land uses.

Dryland crop production is expected to decline significantly by 2080 with production being restricted to Caprivi in the north alone, and that will be on the very margins of viability. Thus, it can be assumed that dryland crop gross output will decline to some 25% of the current levels. Irrigated production will attain more importance, and indeed, it is currently part of an ambitious expansion programme. However, increasing demand for decreasing surface flows particularly in Orange and Fish river systems in the south central broad integrated region will constrain development. These southern irrigation schemes are most viable, and expansion related to the Kavango, Zambezi and Kunene river systems in the northeast and northwest broad integrated regions will tend to be constrained by financial and economic viability problems. It is therefore assumed that despite increasing demand for irrigation development, the value added to GNI by irrigated agriculture will not grow but will remain at 100% of the current levels by 2080.

Before any consideration of adaptation, conditions expected to prevail by 2080 as a result of climate change may result in losses in the direct economic contribution of primary land uses amounting to N$2.5 billion or some 4% of the GNI. The losses are clearly highest with the agricultural sector. It is noteworthy that the estimated losses do not include indirect impacts on the economy that resulted from backward and forward linkages.

DISCUSSION, CONCLUSIONS AND INTERVENTIONS

The predicted losses in tourism and livestock income described above will result from gradual change in climate over the next 70 years. This means that, on average, tourism value might decline at 0.4% per annum, and livestock income might decline at 1% per annum over that period. Land use systems will adapt autonomously at the same time, and extraneous economic factors will play an important part. Given autonomous adaptation, losses in production will tend to be reflected in loss of growth that would take place in the absence of climate change.

In Table 10, it can be seen that losses are clearly highest with the livestock sector and, in particular, with commercial fenced ranching. This is a result of the fragile financial and economic viability of this system, where a small drop in income results in a devastating loss in net income. Table 11 shows these losses per square kilometre.

Table 10: Estimated losses in direct gross national income contributed by agricultural production and natural resource uses in 2009

Sector and tenure	Losses to GNI (N$'000, where N$1:00 = US$0.13)
Agriculture	
Livestock	
State land	0

Communal land	70,000
Commercial land	1,965,000
Total	2,035,000
Crops	
State land	0
Communal land	118,000
Commercial land	19,000
Total	137,000
Total agriculture	
State land	0
Communal land	188,000
Commercial land	1,985,000
Total	2,172,000
Natural resources	
Tourism	
State land	65,000
Communal land	15,000
Commercial land	190,000
Total	270,000
Wildlife (other)	
State land	18,000
Communal land	4,000
Commercial land	35,000
Total	57,000
Wild plants	
State land	0
Communal land	0
Commercial land	0
Total	0
Fish (inland)	
State land	0
Communal land	0
Commercial land	0
Total	0

Total natural resources	
State land	83,000
Communal land	19,000
Commercial land	225,000
Total	327,000

Barnes *et al.*

Barnes *et al. Pastoralism: Research, Policy and Practice* 2012 2:22, doi:10.1186/2041-7136-2-22

Table 11: Estimated losses in direct gross national income contributed per square kilometre by agricultural production and natural resource uses in 2009

Sector and tenure	Losses to GNI (N$/ km2^, where N$1:00 = US$0.13)
Agriculture	
Livestock	
State land	-
Communal land	2.02
Commercial land	55.47
Total	24.70
Crops	
State land	-
Communal land	3.41
Commercial land	0.55
Total	1.67
Total agriculture	
State land	-
Communal land	5.43
Commercial land	56.01
Total agriculture	26.36
Natural resources	
Tourism	
State land	5.26
Communal land	0.44

Commercial land	5.36
Total	3.28
Wildlife (other)	
State land	1.45
Communal land	0.11
Commercial land	0.99
Total	0.69
Wild plants	
State land	-
Communal land	-
Commercial land	-
Total	-
Fish (inland)	
State land	-
Communal land	-
Commercial land	-
Total	-
Total natural resources	
State land	6.71
Communal land	0.55
Commercial land	6.35
Total natural resources	3.96

Barnes *et al.*

Barnes *et al. Pastoralism: Research, Policy and Practice* 2012 2:22, doi:10.1186/2041-7136-2-22

From an economic perspective, pastoralism and wildlife-based tourism, combined at landscape scale and involving common property management, can be sound investments across Namibia's communal lands. Furthermore, as climate changes, the existing economic benefits to the Namibian economy associated with pastoralism and wildlife-based tourism will likely be magnified, as capital-intensive commercial fenced ranching systems become less economic. The inherent flexibility of small-scale livestock keeping, if it can be combined with enhanced community rangeland management, has significant potential. Without adaptation, fenced commercial ranching will likely increasingly fail and will tend to be replaced by less capital-intensive herding systems. In terms of long-term adaptation, it suggests that medium- to large-scale livestock farming systems will tend to become lower input in nature, with systems resembling the cattle posts of the communal lands rather than ranches. Cattle over much of the drier savanna areas may tend to be replaced by small stock.

Given the predicted increase in climate variability in the face of deteriorating environmental conditions, livestock land use will need to become more flexible and able to respond to economic forces. Active interventions should include shifts in livestock and rangeland policy, encouraging the adoption of more resilient systems, which are less rigid and more able to change and adapt. Recent research and analysis in Namibia (Humavindu et al. 2011) has found that adoption of good rangeland management practices, involving flexible herding, economies of scale and commercially viable utilisation of invasive bush, can significantly enhance private profitability and the economic contribution of livestock production. These inherent incentives are constrained by the fact that sound management requires significant skill. Policy as espoused in the new national rangeland management policy (MAWF 2012) recognises this and contains measures to enhance the necessary skills and promote good rangeland management. The community-based rangeland management initiative currently being promoted by the Millennium Challenge Corporation is showing signs of success and is relevant here.

Rainfed cropping, already marginal in terms of viability, will likely become nonviable and will be almost eliminated. Current national policy is ambitious regarding irrigation development, and this drive will likely be strengthened by losses in rainfed cropping. However, expansion of irrigation will likely be curtailed by increasing scarcity of water, particularly in the south, and problems with financial and economic viability, particularly in the north (Turpie et al. 2010). Active intervention should be focused on expansion of irrigation, based on sound development planning, to ensure the development is environmentally and economically efficient.

One example of adaptation already taking place is that, for largely economic reasons, wildlife stocks are currently in the process of increasing, and in certain parts of the country, they are displacing livestock herds. The climate-induced losses in demand for tourism will tend to be offset by natural growth in overall international demand for tourism and the potential for expansion of tourism in underutilised areas, particularly in state-protected areas and communal land. Climate-induced losses in grazing wildlife numbers should be offset to some extent by increases in wildlife on understocked land as well as the taking up of unused capacity for wildlife use in state-protected areas and communal lands. Wild plant use is generally well under the national potential for use of these resources (Barnes et al. 2010), and it is generally likely to increase as other livelihood options through agriculture are curtailed.

Table 10 indicates that income from natural resource use will be more resilient in the face of climate change. This is to be expected, given the generally lower reliance of these activities on primary production and rangeland carrying capacity. It suggests that adaptation to climate change should include strong promotion of natural resource-based land uses. This is something that is already taking place, as a result of economic forces and financial incentives. The CBNRM programme has been very successful in integrating and mainstreaming natural resource use in communal land use (NACSO 2004,2006, 2008). It has been shown to be economically viable (Barnes 2008), has economies of scale and facilitates integrating land uses in a flexible manner under common property management. The commercial land conservancy programme has also introduced economies of scale and spatial flexibility to land use there.

Active intervention should thus involve further expansion of the CBNRM programme. It should be broadened to embrace rangeland management and wild plant use. It should be aimed at maximising comparative advantages in an environmentally and economically efficient way, for example, ensuring that those areas targeted as key in relation to losses in biodiversity are conserved. It should also be carefully planned to maximise its impact on rural livelihoods.

At a national level, implementation of the NRMP will promote improved management of national rangelands, which has been shown to be economically efficient. (Humavindu et al. 2011),show the benefits of 'pincer movement' focused on fewer national level subsidies coupled with both direct incentives to allow stocking and destocking and open access to all markets, and indirect incentives focused on community mobilisation and institution building in communal areas.

Together, these measures harbour the potential to deliver economic benefits of rangeland management at the landscape and community levels at the same time as climate-proofing natural resource land use.

AUTHORS' CONTRIBUTIONS

JIB conducted research, developed models, conducted analysis, and contributed to first drafting of the manuscript. JM coordinated and supervised the work and drafted the final manuscript. MA conducted research, updated models, conducted analysis and drafted the initial manuscript. All authors read and approved the final manuscript.

REFERENCES

1. Alberts, M. 2007. Financial management analysis of park tourism in Namibia. MBA Thesis. Durban, South Africa: Department of Business Studies, Management College of Southern Africa.
2. Barnes, JI. 2008. Community-based tourism and natural resource management in Namibia: local and national economic impacts. In Responsible tourism: critical issues for conservation and development, ed. Spenceley A, 343–357. London: Earthscan.
3. Barnes, J, and B Jones. 2009. Game ranching in Namibia. In Evolution and innovation in wildlife conservation: from parks and game ranches to transfrontier conservation areas, ed. Suich H, Child B, Spenceley A, 113–126. London: Earthscan.
4. Barnes, JI, J MacGregor, and LC Weaver. 2002. Economic efficiency and incentives for change within Namibia's community wildlife use initiatives. World Dev 30 (4): 667–681
5. Barnes, JI, O Nhuleipo, PI Muteyauli, and J MacGregor. 2005. Preliminary economic asset and flow accounts for forest resources in Namibia. DEA Research Discussion Paper No 70, Directorate of Environmental Affairs. Windhoek, Namibia: Ministry of Environment and Tourism.
6. Barnes, JI, J Cannon, and J MacGregor. 2008. Livestock production economics on communal land in Botswana: effects of tenure, scale and subsidies. Development Southern Africa 25 (3): 327–345.

7. Barnes, JI, O Nhuleipo, AC Baker, PI Muteyauli, and V Shigwedha. 2009. Wildlife resource accounts for Namibia, 2004. DEA Research Discussion Paper No 79, Directorate of Environmental Affairs. Windhoek Namibia: Ministry of Environment and Tourism.

8. Barnes, JI, J MacGregor, O Nhuleipo, and PI Muteyauli. 2010. The value of Namibia's forest resources: preliminary economic asset and flow accounts. Development Southern Africa 27 (2): 159–176.

9. Brown, LH. 2009. The effects of climate change on revenues to commercial agriculture and tourism enterprises in Namibia. York, UK: MSc Environmental Economics Thesis, University of York.

10. Humavindu, M, J Barnes, C Nott, N de Klerk, and B Kruger. 2011. Economic valuation of good rangeland management and incentives for efficient implementation for the National Rangeland Strategy. Report to the GEF. Namibia: Ministry of Environment and Tourism.

11. IPPR [Institute for Public Policy Research]. 2009. National accounts 1990–2006 data base.http://www.ippr.org.na (accessed November 2009) PubMed Abstract | Massyn, PJ, E Humphrey, D Grossmann, and J Barnes. 2008. Tourism scoping report: Kunene People's Park. Directorate of Parks and Wildlife Management. Windhoek, Namibia: Ministry of Environment and Tourism.

12. MAWF. 2012. National rangeland management policy (NRMP): restoring Namibia's rangelands. Windhoek, Namibia: Ministry of Agriculture Water and Forestry.

13. Mendelsohn, J, S el Obeid, N de Klerk, and P Vigne. 2006. Farming systems in Namibia. Windhoek, Namibia: Namibia National Farmers Union (NNFU).

14. Midgley, GF, GO Hughes, W Thuiller, G Drew, and W Foden. 2005. Assessment of potential climate change impacts on Namibia's floristic diversity, ecosystem structure and function. Namibian National Biodiversity Programme. Windhoek, Namibia: Directorate of Environmental Affairs.

15. NACSO. 2004. Namibia's communal conservancies: a review of progress and challenges. Windhoek, Namibia: Namibian Association of CBNRM Support Organisations.

16. NACSO. 2006. Namibia's communal conservancies: a review of progress and challenges in 2005. Windhoek, Namibia: Namibian Association of CBNRM Support Organisations.
17. NACSO. 2008. Namibia's communal conservancies: a review of progress and challenges in 2007. Windhoek, Namibia: Namibian Association of CBNRM Support Organisations (NACSO).
18. NTB. 2008. Namibia Tourism Satellite Account -. Windhoek, Namibia: Namibia Tourism Board.
19. Reid, H, L Sahlén, J Stage, and J MacGregor. 2007. The economic impact of climate change in Namibia: how climate change will affect the contribution of Namibia's natural resources to its economy. London: Environmental Economics Programme Discussion Paper 07–02. International Institute for Environment and Development (IIED).
20. Scheiter, S, and SI Higgins. 2009. Impacts of climate change on the vegetation of Africa: an adaptive dynamic vegetation modelling approach. Glob Chang Biol 15: 2224–2246.
21. Sherbourne, R. 2009. Guide to the Namibian economy 2009. Windhoek, Namibia: Institute for Public Policy Research (IPPR).
22. SIAPAC. 2007. Study on characteristics of demand of tourists and use patterns of protected areas in Namibia. Windhoek, Namibia: Ministry of Environment and Tourism.
23. Turpie, J, G Midgley, C Brown, J Barnes, J Pallett, P Desmet, J Tarr, and P Tarr. 2010. Climate change vulnerability and adaptation assessment for Namibia's biodiversity and protected area system. Strengthening the Protected Area Network (SPAN) Project. Windhoek, Namibia: Ministry of Environment and Tourism.
24. WTTC. 2006. Namibia: the impact of travel and tourism on jobs and the economy. Windhoek, Namibia: Namibia Tourism Board and Ministry of Environment and Tourism.

Chapter 9

Systemic Risks of Genetically Modified Crops: the Need for New Approaches to Risk Assessment

Hartmut Meyer

Federation of German Scientists (Vereinigung Deutscher Wissenschaftler, VDW), In den Steinäckern 13, Braunschweig, 38116, Germany

ABSTRACT

Purpose

Since more than 25 years, public dialogues, expert consultations and scientific publications have concluded that a comprehensive assessment of the implications of genetic engineering in agriculture and food production needs to include health, environmental, social and economical aspects, but only very few legal frameworks allow to assess the two latter aspects. This article aims to explain the divergence between societal debate and biosafety legislation and presents approaches to bring both together.

Main Features

The article reviews the development of biosafety regulations in the USA and the EU, focussing on diverging concepts applied for assessing the risks of genetically modified organisms (GMOs).

Results

The dominant environmental risk assessment methodology has been developed to answer basic questions to enable expedient decision making. As a first step, methodologies that take into account complex environmental and landscape aspects should be applied. Expanding the scope of risk assessment, more holistic concepts have been developed, for example the Organisation for Econonomic Co-operation and Development (OECD) concept of systemic risks which includes socio-economic aspects. International bodies as the OECD, the Convention on Biological Diversity (CBD) and the European Union (EU) have developed the Strategic Environmental Assessment (SEA) as an instrument that includes the additional aspects of risk assessment as demanded by many stakeholders. Interestingly, there had been no attempts yet to link the existing frameworks of GMO risk assessment and SEA.

Conclusions

It is recommended to adapt current models of SEA to assess the systemic risks of GMOs. It is also suggested to revise the EU GMO legislation to promote the inclusion of SEA elements.

GENETIC ENGINEERING IN AGRICULTURE: IMPACTS AND RESTRAINTS

The first genetically modified organisms (GMO) deregulated and commercialised was the Flavr Savr tomato in 1994 in the USA, which did not prove to be commercially viable. US genetically modified (GM) agriculture actually started with Bt cotton planting in 1995, but it only was the introduction of Roundup Ready soybeans in 1996, being exported worldwide as basic ingredient for the feed and food industry that initiated the worldwide public debate on the use of GM crops. Meanwhile, James reports that 15 countries grow more than 50,000 ha of GM crops each with a sum of 133.9 million hectares [1]. According to FoEI--pointing to the fact that the data presented by James are mostly based on personal communic1ations by representatives of the biotechnology industry, which also funds his work--this area equates to 9.2% of the arable land worldwide [2]. Ninety-two percent of this area is located in five countries (USA, Brazil, Argentina, India, Canada). GM crop agriculture relies on five plant species (soybean, maize, canola, sugar beet and cotton) predominately producing animal feed, ethanol and fibres in high-input farming systems. Based on the data provided by James, it can be concluded that GM food products mainly comprise sugar, high-fructose corn syrup, soy protein, lecithin or different oils [1]. Some GM maize varieties can be used for direct consumption as, for example, in South Africa. In the USA, some GM papaya is marketed. The range of new properties used in GM crop agriculture is essentially limited to two features: resistance against the herbicides glyphosate and glufosinate and production of Bacillus thuringiensis (Bt) endotoxins that are used to kill specific lepidoptera and coleoptera larvae (Table1).

Table 1: Overview on deregulated and cultivated GM traits in the USA 1992-2009

henotype	Plant species	Number of deregulated traitsa	Transgenic species in cultivationb
Herbicide tolerance		48	
Glyphosate	Canola, cotton, maize, soy, sugar beet	15	Yes
Glufosinate	Canola, cotton, maize, rice, soy, sugar beet	27	Yes, not all species
Others	Cotton, flax, maize, soy	6	Yes, not all species
Insect resistance		41	
Corn borer	Maize	26	Yes
Corn root worm	Maize	3	Yes
Colorado beetle	Potato	11	No
Other	Tomato	1	No
Altered fruit ripening		40	
Flavr Savr	Tomato	33	No
Other	Tomato	7	No
Virus resistance		12	
Papaya ringspot virus	Papaya	3	Yes
Cucumber mosaic virus, zucchini yellow mosaic virus, watermelon mosaic virus 2	Squash	2	Yes
potato leaf roll virus, potato virus Y	Potato	6	No
Plum pox virus	Plum	1	No
Male sterility	Cichoria, canola, maize	8	No
Altered oil composition	Canola, soy	5	No
Higher lysin content	Maize	1	No
Lower nicotine content	Tobacco	1	No

[a]http://www.aphis.usda.gov/biotechnology/not_reg.html, accessed 30 April 2010; [b] [1, 6]

Source: own compilation

Meyer Environmental Sciences Europe 2011 23:7, doi:10.1186/2190-4715-23-7

The main bottleneck for developing a higher variety of commercially viable products seems to be the limited potential of the technology itself. Complex characteristics of plants as drought or saline resistance are based on reactions of the plant organism at several, including but not only the genetic level. Many--still unknown--genes may play a role in the response to environmental condition. The application of genetic engineering alone might not lead to the improvement of such complex traits [3-5]. Only GM plants possessing genes--which are supposed to work in isolation from the plant's metabolism, as the herbicide resistance and Bt genes--are used commercially. Additionally, two GM plant types possessing pathogen-resistant genes which are supposed to interact with an invading organism could be developed into a commercial product: GM virus-resistant papaya and squash grown on 2,000 ha each in the USA [6]. Until the end of 2004-which should leave enough time for the development of commercial seed until 2009-the U.S. authorities approved 877 field trials with plants that were supposed to be virus resistant (988 until the end of 2009). Experiments with GM plants that were supposed to be resistant against fungi did not result in any commercial product yet, 622 field trials were approved in the USA until the end of 2004 (854 until the end of 2009). The main blocks to market fungi-resistant GM plants are the lack of deeper understanding of the molecular plant-fungi interactions and the unsatisfactory levels of resistance [7, 8].

Stein and Rodríguez-Cerezo predict that a turning point has been reached in the limited commercialisation of GM traits [9]. The authors estimate that in 2015 the number of traits in farmers' fields might quadruple to 120, amongst them 17 soy traits (12 herbicide resistant, three altered oil composition, two pest resistant) or 15 rice traits (six insect resistant, four pest resistant, three herbicide resistant, two β-carotene). This development would mainly increase the number of traits mentioned above to 114. Only six traits aim at influencing more complex characteristics as drought resistance in maize while they still rely on single gene alterations.

DEVELOPMENT OF REGULATORY BIOSAFETY FRAMEWORKS

Asilomar Conference

It was U.S. scientists working in the fields of cancer research and molecular biology being concerned about the potential health risks of their work who started the scientific debate on the pros and cons of GMOs [10]. The participants of the 1973 Gordon Conference on Nucleic Acids drafted a resolution, which warned about the potential health risks of hybrid DNA molecules and called successfully upon the National Institutes for Health (NIH) to develop safety guidelines [11]. An international conference to support the development of safety standards was announced and even moratoria on certain types of experiments suggested [12]. In spring 1975, participants of the Asilomar Conference recognised that more than health problems might arise from the industrial, medical and agricultural application of genetic engineering, but they restricted their debates on this risk issue. While the conference concluded that mechanisms of self-control and voluntary guidelines should be the basis for the development of the technology, calls for a stricter and legally binding governmental oversight were launched during the emerging public debate in cities as Cambridge, Massachusetts, harbouring major research institutions [12,13]. Envisaging a growing unease of the public, prominent molecular biologists soon questioned the value of the early risk debate [14-16].

Emerging Biosafety Systems in the USA

When Cohen reported that his research enables scientists to cross the species barriers, suggesting the invention or creation of new species, U.S. politicians started, soon after, to draft regulations for the application of GMOs [17]. This in turn alerted those scientists that envisaged large economic potential based on their work and patents, and in 1977, a draft law for GMO regulation was stalled when Cohen convinced politicians that the results of the new technology could also have appeared in nature. Expecting a revolution in biology and an immense impact on business, genetic engineering was declared as equivalent to conventional breeding methods, meaning a GMO is not a new organism with unforeseeable risks and does not require specific regulation [18]. In 1976, the NIH adopted guidelines, which set up a system based on biological and physical containments. Later, the U.S. National Research Council formalised the risk assessment approach [19]. When in 1983 the first GM bacteria and plants were released in field trials in California, the existing health protection guideline concept was applied to assess possible environmental risks [20]. The U.S. has opted using existing frameworks to set up a consultation system.[1] Nowadays, genes and proteins that render herbicide tolerance to GM plants are assessed and deregulated according to the rules for food additives; plants possessing Bt genes and proteins fall under the pesticide approval rules and growth hormone-producing fish has to be checked under the procedures for approval of animal drugs. Two recent U.S. law cases stated that the procedure agreed upon by the authorities and the applicant for deregulating herbicide-resistant golf lawn grass and alfalfa were faulty. A more rigid assessment under the norms of U.S. environmental laws had to be conducted. With these court decisions it seems that GM plants that can interact substantially with wild or domesticated genetic resources via pollen flow must undergo a more detailed risk assessment in the USA as, for example, GM soy or maize. It remains open until a final Supreme Court decision; if and how these court cases will influence the future GM crop regulation in the USA.

Biosafety Frameworks at the European and UN Level

In contrast to the situation in the USA, the debate in EU countries went beyond expert circles and involved more NGOs and citizen groups. It also lacked the strong focus on emerging commercial prospects of genetic engineering. While the model of the NIH guidelines was adopted by many European governments, the emerging public debate quickly reached the decision that an overarching, specific legal framework was necessary due to the novelty of GMOs [18, 21]. The first biosafety laws were adopted in Denmark in 1986 and Germany in 1990, EU biosafety regulations followed in 1990.[2] Since that time, the concept of the European biosafety legislation is that the properties and behaviour of organisms which "genetic material has been altered in a way that does not occur naturally by mating and/or natural recombination" cannot be predicted from the current experience with and knowledge about the parent organism. Although this so-called process-based system was developed under the umbrella of the community environmental law it did not adapt existing instruments for assessing environmental risks of technical and industrial activities, e.g. environmental impact assessment, but kept the GMO risk assessment approaches that had been developed in the context of the technology development.

In 1995, the negotiations of international binding biosafety rules under the framework of the CBD) started, which resulted in the Cartagena Protocol on Biosafety (CPB)[3] adopted in 2000 [22]. Comparable to the EU, the CPB adopted a process-based type of GMO regulation. As the Biosafety Clearing House of the CPB and other data banks show, legally binding specific biosafety legislation are currently in force or under development in 112 out of 200 countries:

- Seventy-nine states with legislation in force (amongst them 33 industrialised countries)
- Thirt-three states with legislation in development
- Fifty states with a national biosafety framework based on the CPB
- Eleven states having ratified the CPB

Countries, which so far do not follow the process-based approach to biosafety legislation, are the USA and Canada. Twenty-five states have no biosafety system at all.

CONFLICTING CONCEPTS FOR ASSESSING ENVIRONMENTAL RISKS OF GMOS

The "Ecotoxicological Approach" Versus the "Environmental Approach"

Ever since the first GMOs were released, it was discussed whether it is justifiable to apply methods developed for toxicology assessment of chemical substances to viable and reproducible organisms or if new methods had to be developed. The differences between the testing approaches were brought to a wider public when Hilbeck et al. and Losey et al. for the first time showed negative effects of Bt toxins and Bt maize pollen on ecologically relevant non-target organisms in laboratory experiments at a time when Bt crops where already deregulated and cultivated commercially in the USA [23-25]. The U.S. authorities did not require an ecologically oriented laboratory or even field test for the deregulation of Bt cotton in 1995 [26]. The respective risk research and assessments were largely and still are based on ecotoxicological laboratory approaches. Standard protocols and organisms are used due to the good reproducibility of experiments, easy breeding of those organisms and low costs of the work. The two different concepts for GMO risk assessment were named "ecological approach" and "(eco)toxicological approach" [27,28]. According to EFSA, the current arguments and representatives are presented by Andow et al. and Romeis et al. [29-31].

Hilbeck et al. questioned whether the design of these ecotoxicological tests would contribute to assessing the ecological risks of Bt crops [32]. For example, the water flea Daphnia magna was exposed to Bt maize pollen and the measurement of "no effects" was judged as "no risk" although the Bt toxin contained in the pollen will not dissolve in the water and Daphnia cannot eat pollen. Similarly "no effect" results with the earthworm Eisenia fetida were accepted although there was no proof that the worms actually had taken up the toxin in the feeding trials. Apart from questionable test designs, it is known that, for example, the widely used earthworm Eisenia fetidadoes not live in agricultural ecosystems [33]. The criticism on using environmentally irrelevant organisms and ill-designed tests added to the existing uncertainty on how to measure "indirect effects", e.g. the effects of the herbicides used together with herbicide tolerant crops, as demanded by the legal framework, how to deal with the foreseeable EU-wide use of antibiotic marker genes in foodstuff made out of GM crops containing these transgenes and how to evaluate the research work pointing to considerable gene flow in GM canola [34]. It was against this background that the EU environmental council[4] declared the stop of all pending GMO application procedures in 1999 until the EU biosafety regulations had been revised.

DIFFERENT REACTIONS ON THE NEW EU BIOSAFETY FRAMEWORK

This scientific dispute in combination with societal and economic impacts influenced the revision of the EU GMO regulations [35]. The new EU biosafety Directive 2001/18/EC supports the ecological approach and prescribes a more detailed environmental risk assessment (ERA), establishes the precautionary principle as baseline for decision making and also serves as ERA reference for the regulation (EC) 1829/2003 on GM food and feed market approval.[5] The five steps of current risk analysis procedures (hazard identification, exposure assessment, consequences assessment, risk characterization, mitigation options) were accepted as valid for GMOs, but methodologies and interpretations should be adapted to meet the specific features of living organisms and their interactions with the receiving environment [36-39]. Although Directive 2001/18/EC establishes a new framework for ERA prescribing the testing of the GMO as such (not only of the new genes and proteins) or the consideration of the receiving environment (not only some field trial locations as basis for an EU-wide approval), a review of the soil ecotoxicological tests presented in GMO dossiers concluded that they do not reflect the new legal requirements [40]. These authors, in line with Andow and Hilbeck and Snow et al., emphasise that it is crucial not to rely on standard test species only but to choose test species representative of the agro-ecological environments in which the GM plants will be grown [41,42]. A recent EFSA Scientific Opinion elaborates extensively on the issue of species selection that should take into account the "ecological relevance of the species, susceptibility to known or potential stressors, anthropocentric value, testability, exposure pathways" of non-target organisms [29]. Furthermore, experiments with the actual GM crops at different levels of complexity have to be performed as basis for a sound risk assessment [43].

The stated deficits in the GMO dossiers and a series of publications that argue against a wider application of the ecological approach in ERA show that the implications of the new legal framework are seen critical by developers of GM crops and scientists advocating their use. A scientist of Syngenta states that "environmental risk assessment research has often attempted to describe the multitude of potential interactions between transgenic plants and the environment, rather than to test hypotheses that the cultivation of transgenic plants will cause no harm." [28]. The ecological approach obviously supports decision makers against approving GM crops, and ecologists advocate even more research into complex ecological interactions. Raybould addresses not only the methodology of ERA but also the central normative problem in the relationship between risk research and risk assessment: who determines what kind of hypothesis has to be tested, which level of scientific knowledge and certainty is needed before making decisions, and where is the border between "need to know and nice to know".

Developers of GM crops suggest different approaches on how to accelerate the GM crop approval under the new EU system. One basic suggestion of Raybould is that "ecologists must avoid the temptation to test null hypotheses [of no difference between a transgenic plant and a non-transgenic comparator]" but test risk hypotheses on adverse effects of GM crops on environmental goods and processes that need to be protected [28]. With regard to the EU political and legal background, it seems questionable if this approach will lead to the desired outcome. First, the necessary decisions on protection aims have not yet been taken in the EU. Furthermore, the suggestion does not reflect the concept of the EU biosafety legislation saying that the application of gene technologies might lead to new risks and that, therefore, the first requirement of risk assessment is to test the above-noted null hypothesis on unforeseen differences between the GMO and its parents.

The second suggestion of private sector representatives of the ecotoxicological approach is that field tests should not be a prerequisite for GMO approvals, but should only be demanded when literature studies or ecotoxicological experiments show significant negative effects [44]. A scientist of Monsanto suggests that this model should also be applied to his company's drought-resistant GM maize, a trait that until now was seen as model case for more complex, ecologically oriented risk research and assessment [45]. This approach enabling a more expedient approval of GM crops was supported by U.S. and EU governmental risk assessors and public scientists in a joint publication on risk assessment of non-target effects of Bt crops and accordingly shaped the draft guidance on GM crop risk assessment presented by the European Food Safety Authority [31,46].

NORMATIVE DIMENSIONS OF RISK ASSESSMENT

In those discussions, it became apparent that ERA steps 1 and 5 as described by Hill are not restricted to the application of scientific methodology but must also be based on substantial normative and thus value-loaded decisions [37]. Many authors state that step 1 indeed needs to be broadened and developed into a "Problem Formulation". Scientists advocating the ecological approach developed the problem formulation and option assessment (PFOA) tool, based on stock-taking exercises, stakeholder consultation and broader public participation procedures [47]. The PFOA was tested in developing countries not only to improve the ERA but as a technology assessment tool following the suggestion of OECD [48-51]: "Analyses leading to risk management decisions must pay explicit attention to the range of standpoints, in particular in situations with a high potential for controversy. This is often best done by involving the spectrum of participants in every step of the decision-making process, starting with the very formulation of the problem to be analysed. Introducing more public participation into both risk assessment and risk decision-making would make the process more democratic, improve the relevance and quality of technical analysis, and increase the legitimacy and public acceptance of the resulting decisions."

When Raybould reflected on the UK farm scale evaluation of GM herbicide tolerant (GMHT) crops, he illustrated clearly that the problem formulation (step 1) strongly depends on the respective stakeholder interests [52].[6] From a herbicide-producing company's perspective, the preservation of arable weeds presents no value and the aim of any GMHT crop system is to reduce their abundance; from a nature conservation perspective, however, arable weeds are a valuable part of biodiversity that should not be eradicated in agro-ecosystems.

While this attitude of a scientist from the private sector is not very surprising, it can be observed that public scientists in application-oriented fields as plant biotechnology tend to adopt comparable attitudes [53]. Kvakkestad et al. interviewed 62 Scandinavian scientists on their perspectives with regard to the deliberate release of GM crops against their professional and funding backgrounds[54]. Two perspectives prevail: perspective 1 is held by many publicly funded scientists who emphasised that the environmental effects from GM crop are unpredictable, and perspective 2 is held mainly by scientists from the biotechnology industry who emphasise that GM crops present no unique risks. No ecologist associated himself with perspective 2. Publicly funded scientists that do not hold above perspective 1 but promote biosafety systems that establish enabling environments for the adoption of GM crops are meanwhile organised in lobby groups as the Public Research and Regulation Initiative,[7] funded by a former Syngenta manager [55].

Also, step 5 and the activities leading to the final decision involve much more than pure science. Millstone et al. stated that the attitude of authorities to deal "asymmetrically" with research that showed negative effects compared to research that could not show negative effects is interpreted by the public as support of the authorities for the developers of GMOs [56]. The Cartagena. Protocol on Biosafety explicitly refers in its Risk Assessment Annex to this common attitude when it obliges its member states to consider that "lack of scientific knowledge or scientific consensus should not necessarily be interpreted as indicating a particular level of risk, an absence of risk, or an acceptable risk". This formulation had been agreed upon by the negotiators as a way on how to implement the precautionary principle in GMO risk assessment and decision making [57]. To address these normative issues in a democratic and socially acceptable way, new processes are needed, which must secure that the point of view of every stakeholder can have its influence on problem formulation in risk assessment and the final decision making [58,51,60].

BROADER APPROACHES FOR ASSESSING THE IMPLICATIONS OF GM CROP AGRICULTURE

New risk concepts that lay the ground for more holistic approaches to assess risks beyond the traditional scope of GMO legislation have been suggested to solve the above-described disputes. In 2003, the OECD International Futures Programme concluded that the classical risk assessment concepts are not suitable to deal with risks that realise themselves or excerpt their influence in larger spatial and/or temporal dimensions [51]. OECD also suggested that the basis of scientific disciplines has to be broadened because "many risk models assume that a hazard is linked from a well-identified source to a single endpoint in more or less linear fashion. That could well prove a seriously flawed assumption if a number of complex evolving factors are at work". In the GMO context, the OECD model of "Systemic Risks" would include the assessment of socio-economic issues as coexistence, patents on seeds and seed monopolies or induced herbicide resistance in weeds.

The OECD concept of systemic risks and risk governance was quickly taken up by scientists and institutions connected with the bank and insurance sector or concerned about the complex effects of chemical pollutants, pandemics or climate change on human health and public health systems[61-64]. With regard to systemic risks of commercial GM crop agriculture, the German research project "GenEERA" developed methods to improve the current ERA with the aim to support the assessment of socio-economic aspects, specifically focussing on the issue of coexistence. Breckling et al. developed geostatistical models to forecast long-term and regional effects of commercial plantation of GM rape seed that cannot be assessed through experimental approaches [65,66]. Complemented by models that allow to scale-up the effects of climatic, crop cultivation, and population parameter on regional GM rape seed dispersal, the project could show that the plantation of GM rape seed would cause systemic risks [67-69]. Model calculations for regions in Northern Germany showed that due to the persistence of transgenes in the soil seed bank 3 years after GM rape seed cultivation, 90% of the fields brought harvests with a GM content above the 0.9% labelling threshold. After 10 years, this percentage was still 5% [70]. Farmers in the state of Schleswig-Holstein, the main rape seed producing region in Germany, would face major external costs to keep the GM content of their harvest below 0.9% if GM rape seed planting would gradually increase to cover 50% of the acreage within 10 years [71]. The follow-up project GeneRisk[8] will adapt these methodologies for assessing systemic risks to Bt maize, complemented by participative approaches involving local stakeholders (results not published yet).

THE APPLICATION OF SOCIO-ECONOMIC ASSESSMENTS

Besides the issue of coexistence, other socio-economic implications are raised in the GMO debate[72]. Still, most legal frameworks do not allow their inclusion in approval process. Governments of industrialised countries and technology developers argue that the methodologies that need to be applied, which would go beyond the current "science-based" risk assessment, are not harmonised and might lead to intransparent and arbitrary decisions [73,74]. Examples for systematic socio-economic assessments are therefore rare; the available literature was compiled in an online archive recently.[9] During the dispute on whether Mexico as a member of the North American Free Trade Agreement could maintain its ban on GM maize trials from 1998, the Commission for Environmental Cooperation (CEC) of the North American Agreement on Environmental Cooperation conducted an extensive assessment of the implications of GM maize agriculture in Mexico,[10]including a socio-cultural assessment [75]. One of the key findings was that many local and indigenous communities regarded the "presence of any transgenes in maize as an unacceptable risk to their traditional farming practices, and their cultural, symbolic, and spiritual value of maize"[76]. This led to the inclusion of a provision in the Mexican biosafety law that allows for banning the planting of GM maize in regions with traditional maize agriculture. Examples for other legislative or administrative measures that have been taken on the basis of socio-economic considerations are the ban of any activities with GM taro and coffee in the country of Hawai'i[11] and the rejection of an application on GM wine yeast in South Africa[12] with regard to the reservations of indigenous communities or the wine industry, respectively. Only recently, the European Commission published a "Roadmap" on how to integrate socio-economic considerations in the existing legal framework on national basis that should help countries, which are willing to grow GM crops to overcome the approval deadlock in the EU [77][13]. The initiative of the European Commission received mainly negative comments from all stakeholders [78,79]. They pointed out that the paper does not offer a convincing legal, administrative and scientific concept to integrate socio-economic considerations into decision making--as for example laid down by the Dutch GMO

Commission--but simply shifts the contentious issue to the member states [80].

STRATEGIC ENVIRONMENTAL ASSESSMENT AND GM CROP DEVELOPMENT

It is apparent that the assessment of the systemic risks of GM crop agriculture needs a broader set of assessment tools as currently used and prescribed by the legal framework. One--also legally--established tool that might be useful in this context is the Strategic Environmental Assessment (SEA, Appendix). SEA is an internationally recognised approach that allows the assessment of the socio-environmental impacts of policies, programmes and plans. In their pioneering publication, Therivel et al. define SEA as: "the formalized, systematic and comprehensive process of evaluating the environmental effects of a policy, plan or program and its alternatives, [...] and using the findings in publicly accountable decision-making" [81]. SEA have been put into practise in a range of countries and as described by Goodland focus on three main classes of work [82, 83]:

- Policies: legislation and other rules;
- Plans and strategies, including regional and sectoral plans; and
- Programmes or sets of coordinated projects.

In the last decade, several policy processes were initiated to develop and adopt SEA concepts in the field of environmental decision making. In the EU, an SEA Directive came into force in 2001, but in contrast to the general features of SEAs only certain plans and programmes but no policies can be assessed[14]. At the international level, the CBD--following its articles 6b and 14-and the OECD adopted SEA guidelines in the fields of biodiversity-related impact assessment and development cooperation, respectively [84-87]. It is interesting to note, but in the light of the above-described historical development of biosafety regulations, it is not surprising that neither the EU, the CBD nor the OECD includes GMO projects and biosafety policies under the scope of SEAs.

Based on the SEAs undertaken in recent years, many academic analyses on the quality of conduct and content have been published (Chaker 2006; Stoeglehner et al. 2009), but the authors also state that only a very limited amount of work on the effectiveness of SEAs with regard to its ability to influence policy decisions is available [88,89]. This might be caused by the more technical interest of the researchers in SEAs and by the fact that the recommendations of SEAs do not have a legally binding character, which makes it difficult to follow their actual influence on policy decisions. In the context of the GMO discussion, it will be useful to follow the SEA approaches in the field of biofuels. Many countries have started assessments of their biofuel policies and sustainability standards [90, 91]. These standards will influence the trade and use of future biofuels. The envisaged systems of sustainability standards that might also contain exclusion criteria is, to a certain extent, comparable with the approval system of GMOs, and thus the features of their SEAs might also be applicable for biosafety policy and GMO project analysis.

A research of current literature indicates a lack of work on connecting SEA with GMO issues. Authors from Taiwan which is a leading Asian country with regard to the application of SEAs note that while sectoral and spatial planning are covered by SEA, this requirement should also extend to policy issues as "WTO accession, [...] development of biotechnology (e.g., genetically modified food), export of nuclear waste for treatment" [92]. Only Linacre et al., in the context of U.S.-sponsored biotechnology and biosafety capacity building projects, have published a first concept on how to apply SEA to support the adoption of GM crops in developing countries [93]. Since the SEA methodology requires a substantial influence of the public on the final recommendation, the authors see this approach as contentious. They note that "careful consideration needs to be given to how the expert and lay panels are constructed and managed in the qualitative assessment phase" to lead to the desired outcome of the process. With regard to the described basic concepts of SEA, specifically the requirement of an open dialogue without a predetermined outcome, it seems questionable if the concept suggested by Linacre et al. actually reflects the characteristics of an SEA and would lead to a more holistic assessment of systemic risks of GM crops [93].

SUMMARY AND CONCLUSIONS

Since the early times of the development and application of genetic engineering, the scientific and public debate on risks and benefits encompassed a broad range of health, environmental, economic and social issues. It has been concluded in numerous stakeholder rounds and scientific publications that a comprehensive assessment and meaningful consideration of the implications of genetic engineering in all these fields would render more scientific strength and social acceptability to the decision-making process. Despite these debates and recommendations, only very few national legal frameworks and no international instrument obliges governments to include other issues than health and environment in the risk assessment procedure.

Due to the strong linkages between public and private research right from the start of the technology in the USA, the procedures for GMO risk assessment and decision making had been set up to be supportive for the promotion of the technology. Based on the ecotoxicological approach of testing of chemicals, the broader socio-economic issues--as listed above but also more complex ecological concerns as long-term--and food web effects lay beyond the scope of the early GMO risk assessments. The debate on the regulation of genetic engineering in Europe focussed more on issues beyond corporate and economic issues. While this caused the legal character of the EU legislation to be a process-based concept under the environmental law, the previously developed ecotoxicological concept of GMO risk assessment were incorporated into the early EU legislation and the international Cartagena Protocol on Biosafety. Since 1998 until today, it is discussed controversially whether and how to develop traditional GMO risk assessment into a comprehensive environmental risk assessment, taking into account principles and methodologies of environmental and biodiversity research. Based on the published literature, an "ecotoxicological approach" and an "environmental approach" can be characterised. The basic distinction between their proponents is their degree of institutional and educational attachment to the development and marketing of GM crops.

In academic debates and work outside of the GMO field two approaches have emerged that, in combination, might be suitable to make the GMO debate more holistic and the decision framework more responsive to the specific social and economic situations in different countries. Work in the OECD and other fora resulted in the concept of "systemic risks", which has gained popularity in assessing risks in financial, economic and health systems. While it is apparent that at the scientific level the integration of a more holistic approach to the dimensions of GMO risks is feasible and indeed led to first results (e.g. by the projects GenEERA and GeneRisk), the existing official risk assessment and decision-making procedures cannot guarantee an appropriate reflection of these findings. A way forward in integrating the concept of "systemic risks" in GMO decision making could be the application of internationally recognised instruments as the Strategic Environmental Assessment. Guidelines and frameworks have been developed by the OECD, the CBD, and the EU. It is recommended to develop concepts and undertake case studies to test the applicability and usefulness of SEAs to be integrated in biosafety systems that allow for the holistic assessment of systemic risks in agro-biotechnology. The current EU discussions on including socio-economic considerations into GMO decision making offer an opportunity to amend national GMO legislation accordingly. When doing so, the experiences of the ongoing work in assessing biofuel policies and sustainability standards through SEAs should be taken into account.

ACKNOWLEDGEMENTS

The author gratefully acknowledges funding by the German Federal Ministry of Education and Research (BMBF) of the research project "GeneRisk" under grant FKZ: 07VPS14A and the fruitful discussions with members of the Vereinigung Deutscher Wissenschaftler (VDW). The views expressed in this paper are those of the author and do not represent views of the VDW or its members.

REFERENCES

1. James C: Global status of commercialized biotech/GM crops: 2009. ISAAA Brief No. 41. Ithaca: International Service for the Acquisition of Agro-biotech Applications; 2010.

2. FoEI: Who benefits from GM crops?. Amsterdam: Friends of the Earth International; 2010.
3. Flowers TJ: Improving crop salt tolerance. J Exp Bot 2004, 55:307-319.
4. Vinocur B, Altman A: Recent advances in engineering plant tolerance to abiotic stress: achievements and limitations. Curr Opin Biotechnol 2005, 2005:123-132.
5. Visarada KBRS, Meena K, Aruna C, Srujana S, Saikishore N, Seetharama N: Transgenic breeding: perspectives and prospects. Crop Sci 2009, 49:1555-1563.
6. Dymond M, Hurr K: The global status of commercialised genetically modified plants 1 July 2008-31 December 2009. Wellington: MAF Biosecurity New Zealand; 2010.
7. Islam A: Fungus resistant transgenic plants: strategies, progress and lessons learnt. Plant Tissue Cult Biotech 2006, 16:117-138.
8. Stuiver M: Engineering fungal resistance in crops. In Plant biotechnology: current and future applications of genetically modified crops. Edited by Halford N. New York: John Wiley & Sons; 2006:225-239.
9. Stein AJ, Rodríguez-Cerezo E: The global pipeline of new GM crops: implications of asynchronous approval for international trade. European Commission, Joint Research Centre, Institute for Prospective Technology Studies, Sevilla; 2009.
10. Rodgers J: Asilomar revisited. Mosaic 1981, 19-25.
11. Singer M, Soll D: Guidelines for DNA hybrid molecules. Science 1973, 181:1114.
12. Berg P, Baltimore D, Nathans D, Boyer HW, Roblin R, Cohen SN, Watson JD, Davis RW, Weissman S, Hogness DS, Zinder ND: Potential biohazards of recombinant DNA molecules. Proc Natl Acad Sci 1974, 71:2593-2594.
13. Herbig J: Die Gen-Ingenieure. München and Wien: Hanser; 1978.
14. Cohen SN: Recombinant DNA: fact and fiction. Science 1977, 195:654-657.
15. Watson JD: An imaginary monster. Bull At Sci 1977, 33:19-20.

16. Watson JD: Remarks on recombinant DNA. CoEvol Quart Summer 1977, 40-41.
17. Cohen SN: The manipulation of genes. Sci Am 1975, 233:25-32.
18. Wright S: Molecular politics. Developing American and British regulatory policy for genetic engineering, 1972 - 1982. Chicago: The University of Chicago Press; 1994.
19. NRC: Risk assessment in the federal government: managing the process. Washington, DC: National Academies Press; 1983.
20. Suter GW: Ecological risk assessment. Boca Raton: Lewis Publishers; 1993.
21. Commandeur P, Joly PB, Levidow L, Tappeser B, Terragni F: Public debate and regulation of biotechnology in Europe. Biotech Dev Monit 1996, 26:2-9.
22. Meyer H: The Cartagena Protocol on Biosafety. Biotech Dev Monit 2000, 43:2-7.
23. Hilbeck A, Baumgartner M, Fried PM, Bigler F: Effects of transgenic Bacillus thuringiensis corn-fed prey on mortality and development time of immatureChrysoperla carnea (Neuroptera: Chrysopidae). Environ Entymol 1998, 27:480-487.
24. Hilbeck A, Moar WJ, Pusztai-Carey M, Filippini A, Bigler F: Toxicity of Bacillus thuringiensis Cry1Ab toxin to the predator Chrysoperla carnea (Neuroptera: Chrysopidae). Environ Entymol 1998, 27:1255-1263.
25. Losey JE, Rayor LS, Carter ME: Transgenic pollen harms monarch larvae. Nature 1999, 399:214.
26. EPA: Bt cotton confirmatory data and terms and conditions of the amendment. Washington, DC: EPA; 2001.
27. Obrycki J, Losey JE, Taylor OR, Jesse LCH: Transgenic insecticidal corn: beyond insecticidal toxicity to ecological complexity. BioScience 2001, 51:353-361.
28. Raybould A: Ecological versus ecotoxicological methods for assessing the environmental risks of transgenic crops. Plant Sci 2007, 173:589-602.
29. EFSA: Scientific opinion on the assessment of potential impacts of genetically modified plants on non-target organisms. EFSA J 2010.

30. Andow DA, Birch ANE, Dusi AN, Fontes EMG, Hilbeck A, Lang A, Lövei GL, Pires CSS, Sujii ER, Underwood E, Wheatley RE: Non-target and biodiversity risk assessment for genetically modified (GM) crops. Proceedings of 9th International Symposium on the Biosafety of Genetically Modified Organisms: September 2006; Korea 2006, 68-73.

31. Romeis J, Bartsch D, Bigler F, Candolfi MP, Gielkens MMC, Hartley SE, Hellmich RL, Huesing JE, Jepson PC, Layton R, Quemada H, Raybould A, Rose RI, Schiemann J, Sears MK, Shelton AM, Sweet J, Vaituzis Z, Wolt JD: Assessment of risk of insect-resistant transgenic crops to nontarget arthropods. Nature Biotechnol 2008, 26:203-208.

32. Hilbeck A, Meier MS, Raps A: Review on non-target organisms and Bt plants. Report prepared for Greenpeace International, Amsterdam. Zurich: EcoStrat GmbH, Ecological Technology Assessment & Environmental Consulting; 2000.

33. Jänsch S, Amorim MJ, Römbke J: Identification of the ecological requirements of important terrestrial ecotoxicological test species. Environ Rev 2005, 13:51-83.

34. Levidow L, Carr S, Wiel D: Genetically modified crops in the European Union: regulatory conflicts as precautionary opportunities. J Risk Res 2000, 3:189-208.

35. Devos Y, Reheul D, Dewaele D, van Speybroeck L: The interplay between societal concerns and the regulatory frame on GM crops in the European Union. Environ Biosaf Res 2006, 5:127-149.

36. Hill RA, Sendashonga C: General principles for risk assessment of living modified organisms: lessons from chemical risk assessment. Environ Biosaf Res 2003, 2:81-88.

37. Hill RA: Conceptualizing risk assessment methodology for genetically modified organisms. Environ Biosaf Res 2005, 4:67-70.

38. Snow A, Moran-Palma P: Commercialization of transgenic plants: potential ecological risks. BioScience 1997, 47:86-96.

39. Andow DA, Zwahlen C: Assessing environmental risks of transgenic plants.Ecol Lett 2006, 9:196-214.

40. Römbke J, Jänsch S, Meier M, Hilbeck A, Teichmann H, Tappeser B: General recommendations for soil ecotoxicological tests

suitable for the environmental risk assessment of genetically modified plants. Integr Environ Assess Manag 2009, 6:287-300.

41. Andow DA, Hilbeck A: Science-based risk assessment for nontarget effects of transgenic crops. BioScience 2004, 54:637-649.
42. Snow AA, Andow DA, Gepts P, Hallerman EM, Power A, Tiedje JM: Genetically modified organisms and the environment: current status and recommendations. Ecol Appl 2005, 15:377-404.
43. Marvier M, McCreedy MC, Regetz J, Kaveira P: A meta-analysis of effects of Bt cotton and maize on nontarget invertebrates. Science 2007, 316:1475-1477.
44. Garcia-Alonso M, Jakobs E, Raybould A, Nickson TE, Sowig P, Willekens H, van der Kouwe P, Layton R, Amijee F, Fuentes AM: A tiered system for assessing the risk of genetically modified plants to non-target organisms.Environ Biosaf Res 2006, 5:57-65.
45. Nickson TE: Planning environmental risk assessment for genetically modified crops: problem formulation for stress-tolerant crops. Plant Physiol 2008, 147:494-502.
46. EFSA: Guidance on the environmental risk assessment of genetically modified plants. EFSA J 2010.
47. Nelson KC, Banker MJ: Problem formulation and options assessment handbook. St. Paul: University of Minnesota; 2007.
48. Hilbeck A, Andow DA: Environmental risk assessment of genetically modified organisms, volume 1: a case study of Bt maize in Kenya. Wallingford: Cabi Publishing; 2004.
49. HilbeckA, Andow DA, Fontes EMG: Environmental risk assessment of genetically modified organisms, Volume 2: methodologies for assessing Bt cotton in Brazil. Wallingford: Cabi Publishing; 2006.
50. Andow DA, HilbeckA, NguyenVT: Environmental risk assessment of genetically modified organisms, volume 4: challenges and opportunities with Bt cotton in Vietnam. Wallingford: Cabi Publishing; 2008.
51. OECD: Emerging risks in the 21st century--an agenda for action. Paris. 2003.
52. Raybould A: Problem formulation and hypothesis testing for environmental risk assessments of genetically modified crops. Environ Biosaf Res 2006, 5:119-125.

53. Gibbons M: Science's new social contract with society. Nature 1999, 402(Suppl 6761):C81-C84.
54. Kvakkestad V, Gillund F, Kjølberg KA, Vatn A: Scientists' perspectives on the deliberate release of GM crops. Environ Values 2007, 16:79-104.
55. Greef W: The Cartagena Protocol and the future of agbiotech. Nat Biotechnol 2004, 22:811-812.
56. Millstone E, van Zwanenberg P, Marris C, Levidow L, Torgersen H: Science in trade disputes related to potential risks: comparative case studies. European Commission; 2004.
57. Meyer H: The precautionary principle and the Cartagena Protocol on Biosafety: development of a concept. In Biosafety first--holistic approaches to risk and uncertainty in genetic engineering and genetically modified organisms. Edited by Traavik T, Li Ching L. Trondheim: Tapir Academic Press; 2007:469-482.
58. Slovic P: The risk game. J Hazard Mater 2001, 89:17-24.
59. Wynne B: Creating public alienation: expert cultures of risk and ethics on GMOs. Sci Cult 2001, 10:445-481.
60. Committee on Improving Risk Analysis Approaches Used by the U.S. EPA: Science and decisions: advancing risk assessment. Washington, DC. 2009.
61. IRGC: An introduction to the IRGC risk governance framework. Geneva. 2008.
62. Helbing D: Systemic risks in society and economics. In Working Paper 09-12-044. Santa Fe Institute; 2009.
63. Briggs DJ: A framework for integrated environmental health impact assessment of systemic risks. Environ Health 2008, 7:61-78. PubMed Abstract | BioMed Central
64. McMichael A: Environmental change, climate and population health: a challenge for inter-disciplinary research. Environ Health Prev Med 2008, 13:183-186.
65. Breckling B, Laue H, Pehlke H: Remote sensing as a data source to analyse regional implications of genetically modified plants in agriculture--Oilseed rape (Brassica napus) in Northern Germany. Ecol Indicat 2009.
66. Breckling B, Reuter H, Middelhoff U, Glemnitz M, Wurbs A, Schmidt G, Schröder W, Windhorst W: Risk indication of

genetically modified organisms (GMO): modelling environmental exposure and dispersal across different scales. Ecol Indicat 2009.

67. Schmidt G, Schröder W: Regionalisation of climate variability used for modelling the dispersal of genetically modified oil seed rape in Northern Germany. Ecol Indicat 2009.
68. Glemnitz M, Wurbs A, Roth R: Derivation of regional crop sequences as an indicator for potential GMO dispersal on large spatial scales. Ecol Indicat 2009.
69. Middelhoff U, Reiche EW, Windhorst W: An integrative methodology to predict dispersal of genetically modified genotypes in oilseed rape at landscape-level--a study for the region of Schleswig-Holstein, Germany. Ecol Indicat 2009.
70. Reuter H, Schmidt G, Schröder W, Middelhoff U, Pehlke H, Breckling B: Regional distribution of genetically modified organisms (GMOs)-up-scaling the dispersal and persistence potential of herbicide resistant oilseed rape (Brassisca napus). Ecol Indicat 2009.
71. Middelhoff U, Reuter H, Breckling B: GeneTraMP, a spatio-temporal model of the dispersal and persistence of transgenes in feral, volunteer and crop plants of oilseed rape and related species. Ecol Indicat 2009.
72. Otsuka Y: Socioeconomic considerations relevant to the sustainable development, use and control of genetically modified foods. Food Sci Tech 2003, 14:294-318.
73. Gupta A: Framing "biosafety" in an international context: the biosafety protocol negotiations. Cambridge: Harvard University; 1999.
74. Falck-Zepeda JB: Socio-economic considerations, Article 26.1 of the Cartagena Protocol on Biosafety: what are the issues and what is at stake? AgBioForum 2009, 12(1):90-107.
75. Brush S, Chauvet M: Assessment of social and cultural effects associated with transgenic maize production. Quebec: Secretariat of the Commission for Environmental Cooperation; 2004.
76. CEC: Maize and biodiversity. The effects of transgenic maize in Mexico--key findings and recommendations. Quebec. 2004.

77. European Commission: New policy for genetically modified organisms (GMO) cultivation. Brussels: European Commission; 2010.
78. Anonymous: EU GMO proposals draw widespread criticism. Brussels: EurActiv Network; 2010.
79. Chipman A: Fears over Europe's GM crop plan. Nature 2010, 466:542-543.
80. COGEM: Socio-economic aspects of GMOs. Building blocks for an EU sustainability assessment of genetically modified crops. Bilthoven. 2009.
81. Therivel R, Wilson E, Thompson S, Heaney D, Pritchard D: Strategic environmental assessment. London: Earthscan Publishers; 1992.
82. Buckley R: Strategic environmental assessment of policies and plans: legislation and implementation. Impact Assess Proj Apprais 2000, 18:209-215.
83. Goodland R: Strategic environmental assessment and the World Bank Group. Internat J Sustain Dev World Ecol 2005, 12:1-11.
84. CBD: Decision VI/7. Identification, monitoring, indicators and assessments. Guidelines for incorporating biodiversity related issues into environmental-impact-assessment legislation or processes and in strategic impact assessment. Montreal. 2004.
85. CBD: CBD Technical Series No. 26--biodiversity in impact assessment. Background document to CBD Decision VIII/28: voluntary guidelines on biodiversity-inclusive impact assessment. Montreal. 2006.
86. OECD: Applying strategic environmental assessment. Good Practice guidance for development cooperation. Paris. 2006.
87. OECD: Strategic environmental assessment and ecosystem services. Paris. 2008.
88. Chaker A, El-Fad K, Chamas L, Hatjian B: A review of strategic environmental assessment in 12 selected countries. Environ Impact Assess Rev 2006, 26:15-56.
89. Stoeglehner G, Brown AL, Kørnøv LB: SEA and planning: 'ownership' of strategic environmental assessment by the planners is the key to its effectiveness. Impact Assess Proj Apprais 2009, 27:111-120.

90. Gnansounou E: Assessing the sustainability of biofuels: a logic-based model. Energy 2010.

91. Jay S: Strategic environmental assessment for energy production. Energy Policy 2010, 38:3489-3497.

92. Ming-Lone L, Yue-Hwa Y: Development and implementation of strategic environmental assessment in Taiwan. Environ Impact Assess Rev 2004, 24:337-350.

93. Linacre NA, Gaskell J, Rosegrant MW, Falck-Zepeda J, Quemada H, Halsey M, Birner R:Strategic environmental assessments for genetically modified organisms. Impact Assess Proj Apprais 2006, 24:35-43.

94. Abaza H, Bisset R, Sadler B: Environmental impact assessment and strategic environmental assessment: towards an integrated approach. Nairobi: UN Environmental Program; 2004.

Chapter 10

Development of Regional Models of Pinus Radiate Height from GIS Spatial Data Supported with Supplementary Satellite Imagery

Pete Watt[1], Andrew Meredith[1], Chen Yang[1], and Michael S Watt[2]

[1]Indufor Asia-Pacific Ltd, Auckland, New Zealand

[2]Scion, Fendalton, Christchurch, New Zealand

ABSTRACT

Background

A number of data sources currently exist that can provide information on forest plantations at a range of scales over an entire rotation cycle. In particular, LiDAR is quickly becoming the technology of choice for harvest planning and providing local-scale estimates of forest structure. Its application is still limited as repeat annual acquisition at this scale is generally cost prohibitive. Development of temporally updateable models that can accurately project important metrics such as tree height between LiDAR acquisitions would be of considerable use to resource managers. The objective of this research was to develop models of *Pinus radiata* height using GIS spatial data supplemented with RapidEye satellite imagery.

Methods

Multiple regression models were constructed to describe maximum canopy height (H_m) derived from LiDAR at two relatively distant study sites located in Kaingaroa and Tairua forests. A randomised selection of 300 m^2 circular plots was made at both sites and average values of H_m within these plots were used for the modelling. Sources of information used for predicting H_m included stand age and spatial information describing environmental variables and stand productivity. This information was supplemented with spectra and vegetation ratios derived from high resolution RapidEye satellite imagery.

Results

The most robust models of H_m that were developed for both sites included a combination of the crop age obtained from the stand GIS, Site Index (obtained from a GIS surface) and the red-edge vegetation ratio (REVI) The final models of H_m had respective R^2 of 0.99 and 0.94 for the Kaingaroa and Tairua sites. At both sites, stand age was the strongest predictor of H_m. However, the inclusion of REVI from high resolution imagery did add an updatable temporal dimension to the model. Changes in REVI are sensitive to the impacts of abiotic and biotic factors that are not captured by stand age and Site Index.

Conclusion

Applied operationally, this model can be used in a GIS environment to estimate tree height and identify areas of anomalous growth or disturbance caused by wind, snow, fire or disease.

BACKGROUND

Managers of forest resources require detailed information about their crops for planning, management and inventory purposes. Traditionally, forest inventories have been carried out through collection of ground-based measurements describing relevant crop features such as stand density, height and volume. However, as these measurements are usually taken over a limited number of plots, and then extrapolated to entire stands, they do not account for spatial variation across stands. Use of remotely sensed data within forest inventories overcomes this limitation.

Airborne light detection and ranging (LiDAR) has proved to be very useful at characterising the forest canopy in three dimensions. Since the first application of LiDAR in forestry over a decade ago (Nilsson 1996), the technology has been widely used to spatially quantify variation in tree height and crown dimensions at resolutions ranging from the stand level (Hall et al. 2005; Naesset and Bjerknes 2001), plot level (Holmgren et al. 2003; Hyyppä et al. 2001; Lim and Treitz 2004; Popescu et al. 2004) to individual tree level (Chen et al. 2006; Coops et al. 2004; Holmgren and Persson 2004; Persson et al. 2002; Popescu and Zhao 2008; Roberts et al.2005; Yu et al. 2004).

The accuracy of LiDAR-derived estimates is reported to be similar to or better than manual field measurement methods for estimating tree height and volume (Naesset 2002; Watt 2005). Data obtained using LiDAR are now used operationally in Finland, Norway, Sweden, Spain, Chile, and Australia to provide forest estimates at the compartment level (Eid et al. 2004). Due to the cost of flying, application of LiDAR, is usually limited to local scales. While regional scale data is sometimes utilised (Zhao and Popescu 2009), repeat annual acquisition at this scale is generally cost prohibitive (Hudak et al. 2002).

Although not as accurate as LiDAR the use of satellite remote sensing does provide a far more cost effective means of predicting crop height. Numerous studies have shown that multispectral satellite data can be utilised in models to map important forest attributes over the land surface (McRoberts and Tomppo 2007; Donoghue and Watt 2006; Shamsoddini et al. 2013; Ozdemir 2008; Kayitakire et al. 2006). However, there are limitations to using remotely sensed data to predict stand dimensions across the entire rotation. Many studies have shown that relationships with biomass and leaf area index tend to saturate after canopy closure when spectra or indices are utilised as sole predictors of stand dimensions (Donoghue and Watt 2006; le Maire et al. 2011). Indices are often favoured as they are likely to be less sensitive to changes in solar illumination and atmospheric conditions than individual bands. Although it is generally accepted that use of remotely sensed data relies on local calibration, little research has evaluated the robustness of relationships when they are applied across different locations.

Spatial information held in geographic information systems (GIS) is also useful for developing models of tree growth. This approach has considerable merit as forest managers generally have access to GIS and a number of potentially useful predictive variables. Rapid increases in the capability of GIS over recent years have seen the development of spatial surfaces covering a diverse range of climatic (Leathwick and Stephens 1998; Leathwick et al. 2003; Tait et al. 2006) and edaphic variables (Watt and Palmer 2012) that in combination with stand information such as age, can be used to model tree productivity. These environmental surfaces have been successfully used to improve empirical growth models (Dzierzon and Mason 2006) and to develop spatial surfaces describing indices of plantation productivity, at a given age, such as Site Index (Palmer, Hock et al. 2009; Watt et al. 2009; Palmer et al. 2012).

Prediction of forest dimensions using data obtained from GIS surfaces and satellite imagery has been less widely researched. Previous research has shown gains in predictive power when models of important stand dimensions are created using spectral data and stand age (le Maire et al. 2011; Gebreslasie et al. 2010; Zheng et al.

2004), relevant indices of productivity (Gebreslasie et al. 2010) and bioclimatic data (Baccini et al. 2008). The present authors are unaware of any research that has investigated the relative contributions of satellite imagery and all three described components of GIS spatial data in prediction of forest dimensions. The objective of this research was to develop models of *Pinus radiata* height using GIS spatial data supplemented with high resolution satellite imagery.

METHODS

Study Sites

Airborne LiDAR and high resolution (5 m) satellite imagery (RapidEye) were acquired over two forests with contrasting topography (Table 1) located in the North Island of New Zealand (Figure 1). The first dataset was obtained from a 4,000 ha strip located on predominantly flat terrain (Table 1) within Kaingaroa forest in the central North Island. The second dataset included all of Tairua forest, a larger (8,900 ha) undulating area (Table 1) located 160 km to the northwest of Kaingaroa forest (Figure 1). Key LiDAR and RapidEye attributes of both datasets are summarised in Table 2.

Table 1: Variation in LiDAR maximum height, stand age, Site Index, mean REVI, and other key climatic and topographic metrics across the two study sites

Variable	Site	
	Kaingaroa	Tairua
LiDAR maximum height (m)	19.9 (0.63-48.9)	25.0 (0.57-45.7)
Age (years)	13.1 (1–32)	16.5 (1–32)
Site Index (m)	29.5 (25.6-32.5)	30.9 (27.1-34.1)
REVI	0.54 (0.10-0.70)	0.39 (0.02-0.54)
Elevation (m)	537 (440–697)	127 (8–335)
Slope (deg.)	2.35 (0.36-19.1)	13.5 (0.42-38.4)
Mean annual rainfall (mm)	1,545 (1,452-1,704)	2,019 (1,836-2,364)

Mean annual temp. (°C)	10.5 (9.7-11.1)	13.8 (12.7-14.4)

The mean is given followed by the range, in brackets.

Watt et al.

Watt et al. New Zealand Journal of Forestry Science 2013 43:11, doi:10.1186/1179-5395-43-11

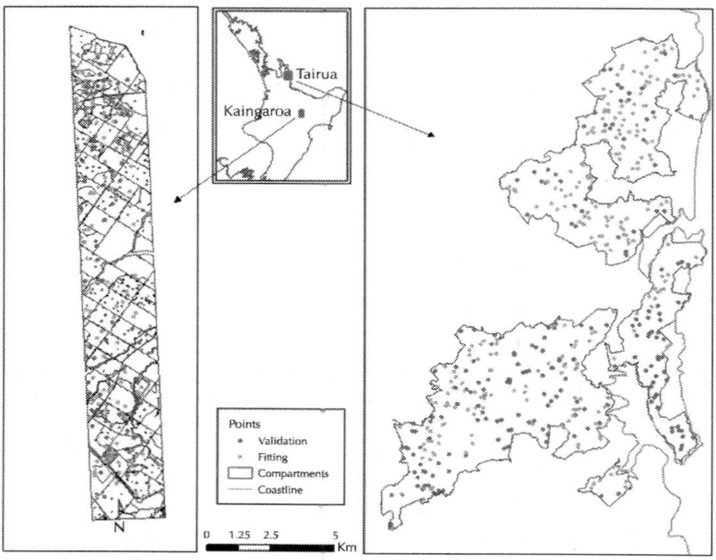

Figure 1: Map showing the location of the plots within both Kaingaroa and Tairua study areas.

Table 2: LiDAR and RapidEye specifications for the two study areas

Variable	Site	
	Kaingaroa	Tairua
LiDAR specifications		
Survey Area (ha)	4000	8900
Survey date	April/ June 2012	July 2012
Pulse density (pulses m-2)	4	2.38
Wavelength (nm)	1064	1064
Scan angle (deg.)	± 6	±6

Pulse frequency (kHz)	70	71
Scan frequency (Hz)	70	53
Footprint diameter (m)	~0.20	~0.20
Ground speed (knots)	105	105
Flying height (m agl)	950	1650
RapidEye specifications		
Image date	27 March 2012	2 May 2012
Resolution (m)	5	5
Spectral bands (No.)	5	5
Sun Angle (deg.)	46	35

Watt et al.

Watt et al. New Zealand Journal of Forestry Science 2013 43:11 doi:10.1186/1179-5395-43-11

LiDAR Dataset

The LiDAR surveys were flown using a fixed-wing aircraft during April/June (Kaingaroa site) and July (Tairua site) 2012 using a small footprint (~0.20 m) Optech ALTM 3100EA system. The LiDAR and flight parameters used to achieve pulse densities of 4 and 2.38 returns m^{-2}, respectively, for the Kaingaroa and Tairua sites are summarised in Table 2. The LiDAR data were processed using customised routines that were developed using Fusion software (McGaughey and Carson 2003). Outputs included a Digital Terrain Model (DTM) and a normalised digital canopy height model. Both were processed to match the spatial resolution of the 5 m RapidEye images. Vertical accuracy surveys based on field data collected in open land-cover areas reported root mean square error (RMSE) values of 0.05 m and 0.11 m, respectively, for the Tairua and Kaingaroa sites.

RapidEye Imagery

Three Rapideye scenes were acquired over the two study sites. Images were selected that were close to the date of the LiDAR survey (see Table 2) with a low level of cloud cover. The quality of each image geo-correction was evaluated against both the LiDAR and stand GIS data and rectifications were made so that the resulting image registration was accurate to a single-pixel resolution (5 m). The impact of topographic shadowing was reduced by applying a topographic correction. This correction adjusts the reflectance of each pixel to a horizontal surface using slope and aspect information extracted from a DTM and the sun angle recorded for each scene. The C-correction method defined in Riano et al. (2003) was applied using a 25 m DTM that was resampled to match the 5 m RapidEye image using bilinear interpolation.

Plot Selection

A randomised selection of 300 m^2 circular LiDAR plots were made across both forests from the LiDAR normalised canopy height model. The forest GIS data were used to select those plots that fell inside *P. radiata* stands across the two study sites. This selection was further refined by discarding plots that fell in unstocked gaps or inside stands that were either irregular in shape or less than five hectares in size. After these exclusions, a total of 247 and 469 plots from the Kaingaroa and Tairua study sites, respectively, were available for analyses. The distribution of these plots and the area covered with the LiDAR is shown in Figure 1.

Data Extraction for Analyses

Maximum canopy height (H_m) for plots was extracted and used as the dependant variable in all analyses. Stand age for each plot was obtained from the stand GIS data.

RapidEye imagery that was spatially coincident with the plots was extracted. From this imagery, reflectance bands were extracted and used in analyses. These reflectance bands were used to determine two different vegetation indices, shown to correlate well with changing densities of canopy leaf area and canopy structure. These included the Normalised Difference Vegetation Index (*NDVI*), determined as,

$$\mathrm{NDVI} = \left(\frac{\rho_{\mathrm{NIR}} - \rho_{\mathrm{Red}}}{\rho_{\mathrm{NIR}} + \rho_{\mathrm{Red}}} \right) \tag{1}$$

where the difference in reflectance from the near infrared (ρ_{NIR}) and red (ρ_{Red}) bands is divided by the sum of the two reflectances. The Normalised Difference Red-Edge Index (*REVI*) takes advantage of the narrow spectral range located between the red-edge reflectance ($\rho_{Rededge}$) and near infrared bands and is determined from,

$$REVI = \left(\frac{\rho_{NIR} - \rho_{Rededge}}{\rho_{NIR} + \rho_{Rededge}} \right) \quad (2)$$

The REVI is more sensitive to changes in chlorophyll concentrations than NDVI. Values for REVI range from -1 to +1 with the index differentiating bare soil at values close to 0 from actively photosynthetic vegetation where values approach 1.

For this study, the band ratios are of particular interest as they provide an effective way to correct for variations in the reflectance that may be caused by differences in atmospheric or sun illumination conditions. This partially negates the need to conduct pre-processing steps such as atmospheric correction.

Environmental data for plot locations were extracted from biophysical GIS surfaces that included primary and secondary terrain attributes (Palmer 2008), biophysical surfaces (Leathwick et al. 2003) and monthly and annual climate data (Mitchell 1991; Leathwick et al. 2002). Key environmental variables used in analyses included mean annual and monthly air temperature, relative humidity, solar radiation, vapour pressure deficit and rainfall. A spatial soil water balance model developed for *P. radiata* (Palmer, Watt et al. 2009) was used to determine mean annual and seasonal root-zone water storage (W) for all plot locations. Fractional available root-zone water storage, W_f, was determined from these data and the maximum available root-zone water storage, W_{max}, as W/W_{max}.

Site Index, which is a standardised measure of height productivity for a site, is defined for *P. radiata* as the mean top height (the average height of the 100 largest diameter trees per hectare) at age 20 years. Values of Site Index for the plots were extracted from a New Zealand surface described in detail in (Palmer, Hock et al. 2009). This surface was developed from Site Index data obtained from an extensive set of permanent sample plots (n = 1,764) established across the environmental range over which *P. radiata* forests occur within New Zealand.

Analyses

All analyses were undertaken using SAS software (SAS-Institute-Inc. 2000). Two separate models of H_m were developed for each of the study sites. Approximately two-thirds of the data was used for model fitting (n = 165 plots for Kaingaroa; n = 312 plots for Tairua) while the remaining one-third was set aside for model validation (n = 82 plots for Kaingaroa; n = 157 plots for Tairua).

For both models, a multiple regression approach was taken to predict H_m from age, Site Index, environmental data and RapidEye spectra and indices. Variables were introduced sequentially into each model starting with the variable that exhibited the strongest correlation until further additions were not significant or did not substantially improve the model performance. Variable selection was undertaken manually, one variable at a time. Plots of residuals were examined prior to variable addition to ensure that the variable was included in the model using the least biased functional form.

Accuracy and bias, were assessed on both the fitting and validation datasets for both models. To assess bias predicted values for H_m were plotted against actual values and residual values were plotted against all variables in the model. Model accuracy was determined through examining the coefficient of determination (R^2) and RMSE.

A further check on the model was made by differencing the LiDAR height surface (which is used as the benchmark) from the surface predicted using the final model of H_m for both study sites. The difference surface was generated at the native resolution of the RapidEye image. Two surfaces were created that included the absolute difference between the LiDAR and the predictive surface and this absolute difference expressed as a percentage of the LiDAR height. Spatial predictions ofH_m excluded recently harvested stands and stands with species other than *P. radiata*.

RESULTS

Data Range

Maximum height obtained from the LiDAR ranged widely within both datasets from ca. 1 m in both study sites to 49 m at the Kaingaroa site and 46 m at the Tairua site (Table 1). Age ranged from 1 to 32 years at both sites. Although mean Site Index was slightly higher at the Tairua than the Kaingaroa site, the range in Site Index was relatively similar between sites at *ca.* 7 m (Table 1). Mean values for REVI were higher at the Kaingaroa site than the Tairua site. The mean annual rainfall at both sites was similar although Tairua was substantially warmer than the Kaingaroa site. The terrain at the Tairua site ranged from flat to very steep and was considerably steeper on average than the relatively flat terrain at the Kaingaroa site (mean slope of 13.5 vs. 2.35°). The Kaingaroa site was located at a considerably higher elevation than the Tairua site (Table 1).

Predictive Models

Kaingaroa

At the Kaingaroa site, maximum LiDAR height was most strongly correlated with tree age (Figure 2a). Age accounted for 98% of the variance in H_m and the RMSE of a model with only this variable was 1.73 m (Table 3). Combining age with Site Index reduced the RMSE to 1.46 m while a model with age and mean REVI had an RMSE of 1.52 m (Table 3). The final model constructed using Age, Site Index and mean REVI from the fitting dataset had an *R2* value of 0.991 and RMSE of 1.29 m (Table 3). For the final model, values of H_m showed little bias when plotted against predictions (Figure 3a) or when residual values were plotted against any of the variables in the dataset (data not shown). Models were similarly ranked using the validation dataset, with the final model having an *R2* value of 0.990 and RMSE of 1.37 m (Table 3). Using the validation dataset there was little apparent bias when predictions were plotted against actual values (Figure 3c).

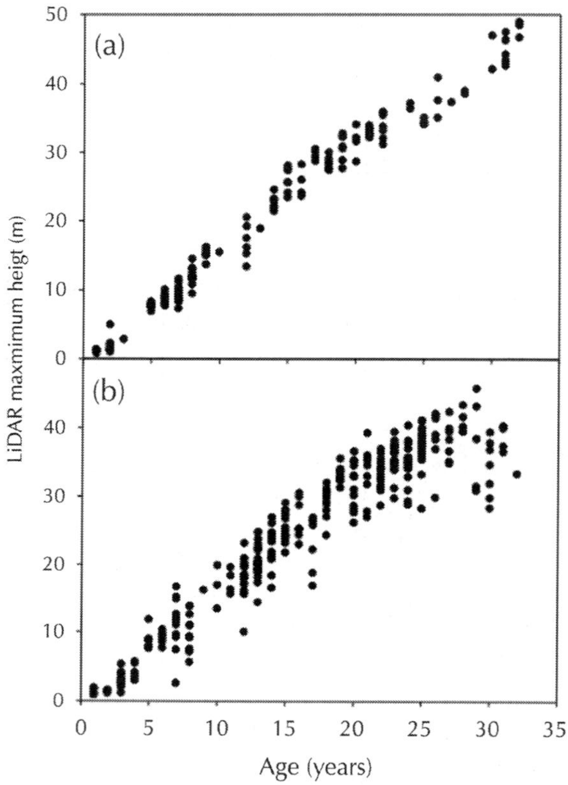

Figure 2: Relationship between LiDAR maximum height and age. The relationship is shown for (a) Kaingaroa and (b) Tairua forests.

Table 3: Summary of statistics for the four regression models of maximum LiDAR height developed using data from Kaingaroa forest

Model no	Variables included	Fitting dataset		Validation dataset	
		R2	RMSE (m)	R2	RMSE (m)
1	Age	0.984	1.73	0.982	1.83
2	Age, Site Index	0.989	1.46	0.986	1.61
3	Age, Mean REVI	0.988	1.52	0.988	1.52
4	Age, Site Index, Mean REVI	0.991	1.29	0.990	1.37

Shown are the coefficient of determination (R^2) and root mean square error (RMSE) for the fitting and validation datasets. All variables in the four models were significant at $P<0.001$.

Partial response functions for the final model show that H_m was most sensitive to changes in age, with values of H_m increasing 45.3 m across the age range in the dataset (Figure 4a). There was considerably less sensitivity between H_m and Site Index with values of H_m increasing 2.8 m across the Site Index range (Figure 4b). The relationship between H_m and mean REVI was characterised by an upward opening polynomial with values of H_m ranging 5.1 m across the mean REVI range and reaching minimum values at a mean REVI value of 0.44 (Figure 4c).

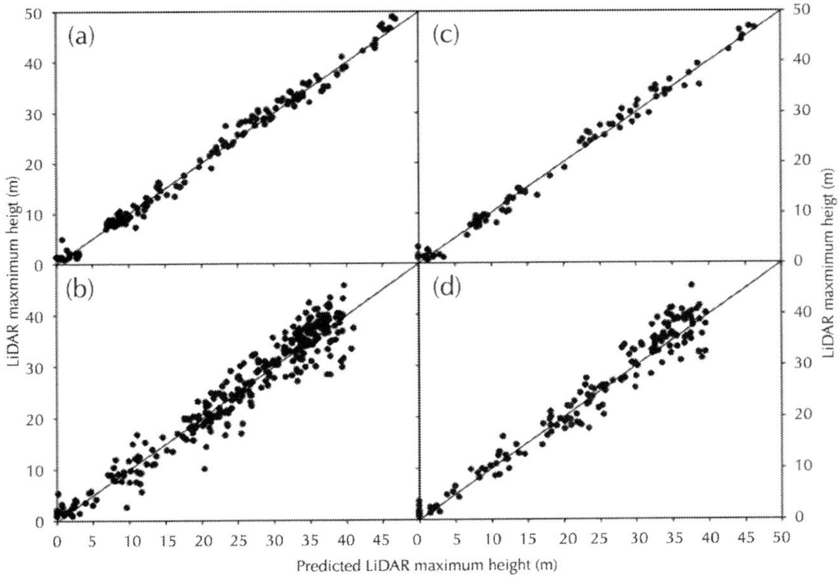

Figure 3: Relationship between actual and predicted LiDAR maximum height. The relationship is shown using data from the fitting dataset for (a) Kaingaroa and (b) Tairua sites and using the validation dataset for (c) Kaingaroa and (d) Tairua sites.

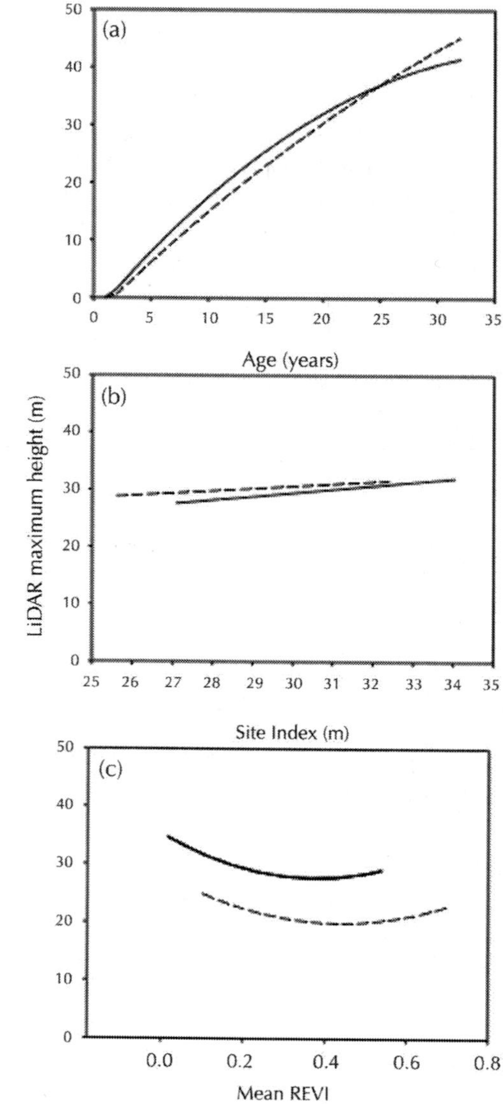

Figure 4: Partial response functions describing the response of predicted LiDAR maximum height. Response functions are shown for(a) age, (b) Site Index and (c) mean REVI, for Kaingaroa (dashed line) and Tairua (solid line) study sites. For each partial response curve, apart from Site Index, all other variables were held at mean values when the response curve was generated across the dataset range for each variable. For Site Index, age was held at 20 years as this is the age at which Site Index is determined.

Tairua

Tree age was the strongest determinant of H_m at the Tairua site accounting for 93% of the variance in the dataset (Figure 2b) with RMSE of 3.13 m (Table 4). Inclusion of either Site Index or mean REVI, with tree age, reduced the RMSE to 3.02 m and 3.07 m, respectively (Table 4). The final model with age, Site Index and mean REVI had an RMSE of 2.96 m and R^2 of 0.936 (Table 4). There was little apparent bias in the final model when predicted values were plotted against actual values (Figure 3b) or residual values were plotted against the variables in the final model (data not shown).

Table 4: Summary of statistics for the four regression models of maximum LiDAR height developed using data from Tairua Forest

Model no	Variables included	Fitting dataset		Validation dataset	
		R2	RMSE (m)	R2	RMSE (m)
1	Age	0.929	3.13	0.951	2.66
2	Age, Site Index	0.934	3.02	0.953	2.60
3	Age, Mean REVI	0.931	3.07	0.948	2.75
4	Age, Site Index, Mean REVI	0.936	2.96	0.951	2.67

Shown are the coefficient of determination (R2) and root mean square error (RMSE) for the fitting and validation datasets. All variables in the four models were significant at P <0.001.

Watt et al.

Watt et al. New Zealand Journal of Forestry Science 2013 43:11, doi:10.1186/1179-5395-43-11

When fitted to the validation dataset, the most precise model included only the variables of age and Site Index. The full model, developed using the fitting dataset, had a similar precision to the earlier model that included only age as a variable. Using the validation dataset, there was little apparent bias when predictions were plotted against actual values (Figure 3d).

Maximum LiDAR height was most sensitive to age with H_m ranging 41.5 m across the dataset (Figure 4a). The impact of Site Index on H_m was not as marked with values of H_m ranging 4.4 m across the Site Index range (Figure 4b). Similarly, there was little sensitivity to mean REVI, with a range of 7.0 m occurring across the range in mean REVI (Figure 4c). The relationship between H_m and mean REVI was characterised by an upward opening polynomial with minimum H_m reached at mean REVI of 0.38.

Predictive Surfaces

For both forests, surfaces of H_m predicted using the final models (Figures 5b and 6b) corresponded reasonably well to H_m derived from the LiDAR (Figures 5a and6a). Surfaces showed absolute differences between the two surfaces to be less than 6 m or 30% in most areas within both sites (Figure 5c, d and 6c, d). At the Tairua site the large percentage differences were often associated with stands with low H_m derived from the LiDAR (areas shaded red in the bottom left of Figure 6d). In these areas absolute differences between modelled and LiDAR derived were relatively low (Figure 6c).

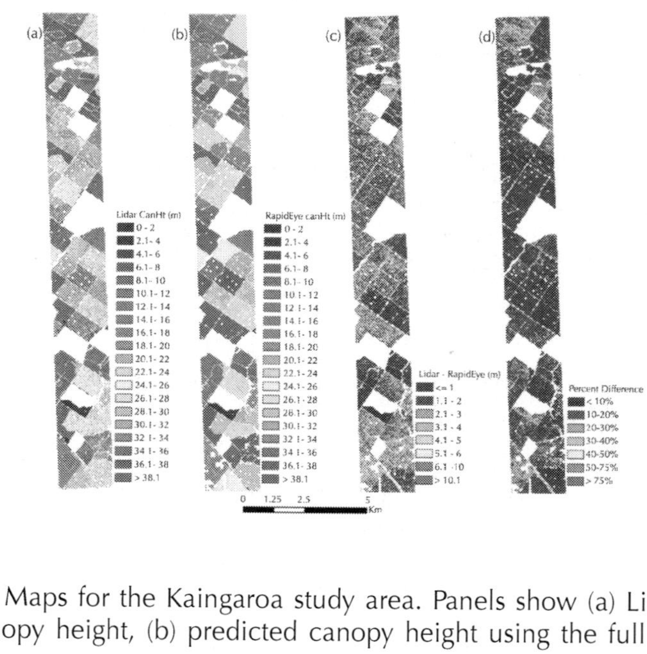

Figure 5: Maps for the Kaingaroa study area. Panels show (a) LiDAR maximum canopy height, (b) predicted canopy height using the full model, (c)

absolute differences between LiDAR and predicted height, and (d) absolute differences expressed as a percentage of the LiDAR height.

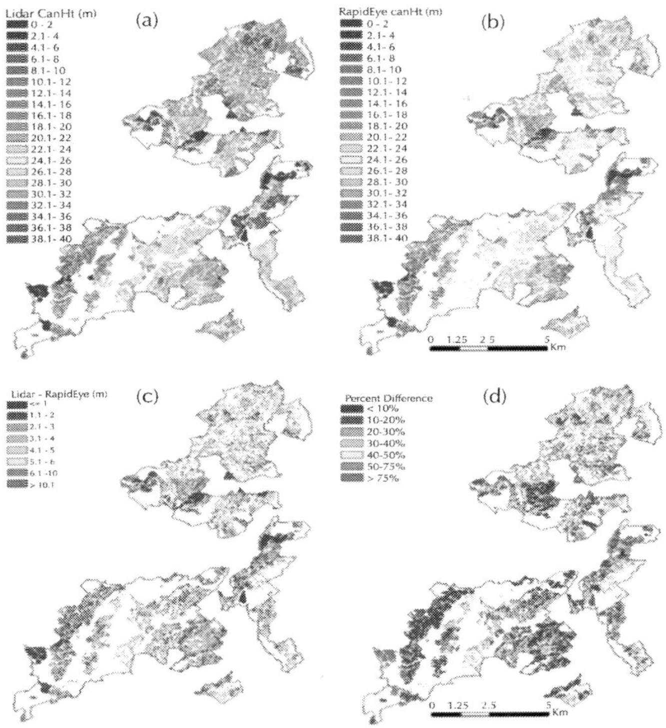

Figure 6: Maps for the Tairua study area. Panels show (a) LiDAR maximum canopy height, (b) predicted canopy height using the full model,(c) absolute differences between LiDAR and predicted height, and (d) absolute differences expressed as a percentage of the LiDAR height.

DISCUSSION

This study clearly showed that precise models of tree height can be developed using stand age and a GIS surface describing Site Index. Supplemental information describing REVI, obtained from fine resolution satellite imagery, significantly improved the models but this information was of secondary importance to use of GIS layers. Although of secondary importance, use of REVI does provide a temporally updateable component to the height models that could potentially account for the impact of abiotic and biotic factors on tree height.

At both study sites, age was the strongest determinant of H_m. Comparison of partial response functions show that age exceeded the contribution of other variables by at least nine-fold at Kaingaroa and six-fold at Tairua. The strength of the age–height relationship was largely attributable to the relatively narrow range in Site Index at both sites, which varied by *ca.* 7 m. Site Index often varies more widely at a forest level, and even more widely at the national level, with an almost three-fold range (from 15.1 to 42.5 m). The relationship between LiDAR height and age will be weaker in forests with greater variability in Site Index.

Inclusion of Site Index in the model allows for more generalised prediction. Empirical height growth models typically use Site Index to scale age-height relationships between sites with differing productivity. Consequently, from a theoretical point of view empirical selection of this variable in the model was sound and reinforced the accuracy of the Site Index layer that was primarily based on productivity relationships with air temperature and rainfall (Palmer, Watt et al. 2009). Age and Site Index should be included within height productivity models given the critical importance of these two variables. In situations where Site Index is not available, use of key environmental variables (such as air temperature and mean annual rainfall) may provide useful surrogates for Site Index.

After accounting for age and Site Index, the strongest correlation was between REVI and H_m. The REVI uses a narrow spectral band width along the red edge band that is sensitive to changes in stress and chlorophyll concentrations of different types of vegetation (Chappelle et al. 1992; Gitelson et al. 1996; Carter and Knapp2001). Previous research using an aerial platform has also identified the red-edge band as providing useful information on photosynthetic vigour (Curran et al.1990).

Gains in precision were greatest when REVI was combined with both age and Site Index. Within the fitting dataset, relatively substantial gains were obtained over the age-only model at both Kaingaroa (RMSE = 1.29 *vs.* 1.73 m) and to a lesser extent at the Tairua site (RMSE = 2.96 *vs.* 3.13 m). The validation process confirmed the importance of REVI at the Kaingaroa site. However gains obtained through adding

REVI and Site Index to the base model with age did not result in model improvement at the Tairua site. This is possibly attributable to increased presence of topographic shadow due to the lower sun angle (35°) and undulating terrain over the Tairua site. These factors effectively mitigate the shadow correction process in areas of high relief. This is evident in site differences in REVI where REVI was markedly lower over the Tairua site.

The functional form of the models was similar for both study sites, despite site differences in the magnitude of REVI. This result provides an indication that the same factors influence the relationship at each site. Previous research has shown that REVI is prone to saturation once the canopy has closed (Donoghue and Watt 2006; le Maire et al. 2011). This study demonstrates utility of REVI at ages beyond canopy closure if the variable is combined with age and other spatially derived layers that describe environmental processes and interactions.

Inclusion of satellite-based metrics, such as REVI, in height models allows for fine-scale spatial predictions that can be temporally updated. For example, within stand variation in productivity is not accounted for by age as this is a compartment-level variable. Similarly, while Site Index partially accounts for productivity variation within stands (at a scale of 100 m^2) this variable is a static measure that is not sensitive to inter- annual variation in site conditions. In contrast, REVI is available at spatial resolutions that allow for fine-scale predictions of productivity. As the REVI layer can be continuously updated, predictions from repeat acquisitions are sensitive to changes in forest structure caused by growth or changes to the forest canopy caused by wind, snow, fire or disease (Watt and Watt2011).

The use of a difference index or ratios such as REVI are of particular interest as they provide an effective way to correct for variations in the reflectance values that may be caused by differences in atmospheric or sun illumination conditions. This partially negates the need to conduct pre-processing steps such as atmospheric corrections. Findings from the current study show that the functional form between REVI and height was very similar between sites although the satellite information used was acquired at different times. These results reinforce the utility of REVI as a general predictor of height growth using data acquired under different atmospheric conditions.

In contrast to the results presented here, previous research has shown that mean height of *P. radiata* can be accurately predicted using only high resolution satellite imagery data (Shamsoddini et al. 2013). Shamsoddini et al. (2013) show strong correlations between mean height and reflectance bands (R^2 of 0.82) and band ratios (R^2 of 0.78) from a WorldView-2 multispectral image. The most precise model of mean height was developed using textural attributes and had a coefficient of determination of 0.93 (Shamsoddini et al. 2013). The approach used by Shamsoddini et al. (2013) is not reliant on up-to-date GIS records, that include recently harvested areas, as the relationships used for prediction are more sensitive to variation between bare ground and forested areas, than those used here that were primarily reliant on stand age derived from the GIS. The approach used in the present study is useful in well managed plantations where managers have access to spatial surfaces describing variation in site productivity such as Site Index. Although the approach outlined here is reliant on up to date stand records a methodology has been developed using RapidEye imagery to update recently harvested areas and segregate different species within the stand GIS (Watt and Watt 2011; Watt and Watt 2012). This methodology could be used to update the stand GIS immediately before the models described in this paper are applied to estimate height. The use of a single variable such as REVI from satellite imagery is justified as the simplicity of the relationship may allow generalisation that is not possible with more complex models that include a number of independent variables.

Further research should be undertaken across broader spatial scales that include greater gradients in Site Index to further test the generality of the models developed here. Development of generally applicable regional models of height would be of considerable use to the forestry sector as this approach provides a means of more frequently mapping and monitoring of crop performance. Once this benchmark has been established then subsequent acquisition of imagery should be able to identify clearfell areas not updated in the stand GIS or areas with abiotic or biotic damage. This information would be useful to both resource planners and managers. Frequent coverage from satellite-based systems will also assist in the focussing of LiDAR surveys around more heterogeneous areas (Watt and Watt2012).

CONCLUSIONS

In conclusion, the retrieval of forest structure information over the lifetime of a planted forest is best achieved by the integration of information from GIS layers with satellite, airborne and ground-based measurements. Through such integration of data into a GIS, foresters have a powerful tool that is capable of providing a rapid overview of a large forest estate. These data provide both a pictorial view of the forest and quantitative analysis that can underpin better informed forest management decisions.

AUTHORS' CONTRIBUTIONS

PW was the primary author. AM and CY extracted the data and produced the maps. MSW undertook the analysis and assisted with the writing. All authors read and approved the final manuscript.

ACKNOWLEDGEMENTS

This project was funded within the Intensive Forest Systems project of Future Forests Research Ltd. We thank two anonymous reviewers for providing useful comments that were used to improve the manuscript.

REFERENCES

1. Baccini, A, Laporte, N, Goetz, SJ, Sun, M, Dong, H (2008). A first map of tropical Africa's above-ground biomass derived from satellite imagery. *Environmental Research Letters, 3*(4).
2. Carter, GA, & Knapp, AK (2001). Leaf optical properties in higher plants: linking spectral characteristics to stress and chlorophyll concentration. *American Journal of Botany, 88*(4), 677–684.
3. Chappelle, EW, Kim, MS, Mcmurtrey, JE (1992). Ration analysis of reflectance spectra (RARS): an algorithm for the remote estimation of the concentrations of chlorophyll a, chlorophyll B, and carotenoids in soybean leaves. *Remote Sensing and Environment, 39*, 239–247.

4. Chen, Q, Baldocchi, D, Gong, P, Kelly, M (2006). Isolating individual trees in a savanna woodland using small footprint lidar data. *Photogrammetric Engineering and Remote Sensing, 72*, 923–932.
5. Coops, NC, Wulder, MA, Culvenor, DS, St-Onge, B (2004). Comparison of forest attributes extracted from fine spatial resolution multispectral and lidar data.*Canadian Journal Of Remote Sensing, 30*, 855–866.
6. Curran, PJ, Dungan, JL, Gholz, HL (1990). Exploring the relationship between reflectance Red edge and chlorophyll content in slash pine. *Tree Physiology, 7*(1–4), 33–48.
7. Donoghue, DNM, & Watt, PJ (2006). Using LiDAR to compare forest height estimates from IKONOS and landsat ETM+ data in Sitka spruce plantation forests.*International Journal Of Remote Sensing, 27*(11), 2161–2175.
8. Dzierzon, H, & Mason, EG (2006). Towards a nationwide growth and yield model for radiata pine plantations in New Zealand. *Canadian Journal of Forest Research,36*, 2533–2543.
9. Eid, T, Gobakken, T, Naesset, E (2004). Comparing stand inventories for large areas based on photo-interpretation and laser scanning by means of cost-plus-loss analyses. *Scandinavian Journal of Forest Research, 19*(6), 512–523.
10. Gebreslasie, MT, Ahmed, FB, van Aardt, JAN (2010). Predicting forest structural attributes using ancillary data and ASTER satellite data. *International Journal Of Applied Earth Observation And Geoinformation, 12*, S23–S26.
11. Gitelson, AA, Kaufman, YJ, Merzlyak, MN (1996). Use of a green channel in remote sensing of global vegetation from EOS-MODIS. *Remote Sensing Of Environment,58*(3), 289–298.
12. Hall, SA, Burke, IC, Box, DO, Kaufmann, MR, Stoker, JM (2005). Estimating stand structure using discrete-return lidar: an example from low density, fire prone ponderosa pine forests. *Forest Ecology and Management, 208*(1–3), 189–209.
13. Holmgren, J, Nilsson, M, Olsson, H (2003). Estimation of tree height and stem volume on plots using airborne laser scanning. *Forest Science, 49*, 419–428.

14. Holmgren, J, & Persson, A (2004). Identifying species of individual trees using airborne laser scanner. *Remote Sensing Of Environment, 90*(4), 415–423.
15. Hudak, AT, Lefsky, MA, Cohen, WB, Berterretche, M (2002). Integration of lidar and landsat ETM plus data for estimating and mapping forest canopy height. *Remote Sensing Of Environment, 82*(2–3), 397–416.
16. Hyyppä, J, Kelle, O, Lehikoinen, M, Inkinen, M (2001). A segmentation-based method to retrieve stem volume estimates from 3-D tree height models produced by laser scanners. *Ieee Transactions on Geoscience and Remote Sensing, 39*, 969–975.
17. Kayitakire, F, Hamel, C, Defourny, P (2006). Retrieving forest structure variables based on image texture analysis and IKONOS-2 imagery. *Remote Sensing Of Environment, 102*(3–4), 390–401.
18. le Maire, G, Marsden, C, Nouvellon, Y, Grinand, C, Hakamada, R, Stape, JL, et al. (2011). MODIS NDVI time-series allow the monitoring of eucalyptus plantation biomass. *Remote Sensing Of Environment, 115*(10), 2613–2625.
19. Leathwick, J, Morgan, F, Wilson, G, Rutledge, D, McLeod, M, Johnston, K (2002). Land environments of New Zealand: a technical guide. (p. 184). Hamilton: Ministry for the Environment, Wellington, and Manaaki Whenua Landcare Research.
20. Leathwick, J, Wilson, G, Rutledge, D, Wardle, P, Morgan, F, Johnston, K, et al. (2003). Land environments of New Zealand. Hamilton: Ministry for the Environment, Wellington, and Manaaki Whenua Landcare Research.
21. Leathwick, JR, & Stephens, RTT (1998). Climate surfaces for New Zealand. (p. 19). Lincoln, New Zealand: Landcare Research.
22. Lim, KS, & Treitz, PM (2004). Estimation of aboveground forest biomass from airborne discrete return laser scanner data using canopy-based quantile estimators.*Scandinavian Journal of Forest Research, 19*, 558–570.
23. McGaughey, RJ, & Carson, WW (2003). Fusing LIDAR data, photographs, and other data using 2D and 3D visualization techniques. (pp. 16–24). Charleston, South Carolina: Bethesda, MD: American Society for Photogrammetry and Remote Sensing.

24. McRoberts, RE, & Tomppo, E (2007). Remote sensing support for national forest inventories. *Remote Sensing Of Environment, 110*, 412–419.
25. Mitchell, ND (1991). The derivation of climate surfaces for New Zealand, and their application to the bioclimatic analysis of the distribution of kauri (*Agathis australis*). *Journal of the Royal Society of New Zealand, 21*, 13–24.
26. Naesset, E (2002). Predicting forest stand characteristics with airborne scanning laser using a practical two-stage procedure and field data. *Remote Sensing Of Environment, 80*(1), 88–99.
27. Naesset, E, & Bjerknes, KO (2001). Estimating tree heights and number of stems in young forest stands using airborne laser scanner data. *Remote Sensing Of Environment, 78*(3), 328–340.
28. Nilsson, M (1996). Estimation of tree heights and stand volume using an airborne LiDAR system. *Remote Sensing Of Environment, 56*(1), 1–7.
29. Ozdemir, I (2008). Estimating stem volume by tree crown area and tree shadow area extracted from pan-sharpened quickbird imagery in open Crimean juniper forests. *International Journal Of Remote Sensing, 29*(19), 5643–5655.
30. Palmer, D, Hock, B, Kimberley, M, Watt, M, Lowe, D, Payn, T (2009). Comparison of spatial prediction techniques for developing Pinus radiata productivity surfaces across New Zealand. *Forest Ecology and Management, 258*(9), 2046–2055.
31. Palmer, DJ (2008). Development of national extent terrain attributes (TANZ), soil water balance surfaces (SWatBal), and environmental surfaces, and their application for spatial modelling of Pinus radiata productivity across New Zealand. (p. 410). Hamilton, New Zealand: PhD thesis, University of Waikato.
32. Palmer, DJ, Watt, MS, Hock, BK, Lowe, DJ (2009). A dynamic framework for spatial modelling *Pinus radiata* soil water balance (SWatBal) across New Zealand. *Scion Bulletin, 234*, 93.
33. Palmer, DJ, Watt, MS, Kimberley, MO, Dungey, HS (2012). Predicting the spatial distribution of *Sequoia sempervirens* productivity in New Zealand. *New Zealand Journal of Forestry Science, 42*, 81–89.

34. Persson, A, Holmgren, J, Soderman, U (2002). Detecting and measuring individual trees using an airborne laser scanner. *Photogrammetric Engineering And Remote Sensing, 68*(9), 925–932.

35. Popescu, SC, Wynne, RH, Scrivani, JA (2004). Fusion of small-footprint lidar and multispectral data to estimate plot-level volume and biomass in deciduous and pine forests in Virginia, USA. *Forest Science, 50*, 551–565.

36. Popescu, SC, & Zhao, K (2008). A voxel-based lidar method for estimating crown base height for deciduous and pine trees. *Remote Sensing Of Environment, 112*, 767–781.

37. Riano, D, Chuvieco, E, Salas, J, Aguado, I (2003). Assessment of different topographic corrections in Landsat TM data for mapping vegetation types. *IEEE Trans. Geosci. Remote Sens, 41*, 1056–1061.

38. Roberts, SD, Dean, TJ, Evans, DL, McCombs, JW, Harrington, RL, Glass, PA (2005). Estimating individual tree leaf area in loblolly pine plantations using LiDAR-derived measurements of height and crown dimensions. *Forest Ecology and Management, 213*(1–3), 54–70.

39. SAS-Institute-Inc (2000). SAS/STAT User's guide: version 8. (p. 3884). Cary, North Carolina: SAS Institute Inc.

40. Shamsoddini, A, Trinder, JC, Turner, R (2013). Pine plantation structure mapping using WorldView-2 multispectral image. *International Journal Of Remote Sensing,34*(11), 3986–4007.

41. Tait, A, Henderson, R, Turner, R, Zheng, Z (2006). Thin plate smoothing interpolation of daily rainfall for New Zealand using a climatological rainfall surface.*International Journal of Climatology, 26*, 2097–2115.

42. Watt, MS, & Palmer, DJ (2012). Use of regression kriging to develop a carbon:nitrogen ratio surface for New Zealand. *Geoderma, 183*, 49–57.

43. Watt, MS, Palmer, DJ, Dungey, H, Kimberley, MO (2009). Predicting the spatial distribution of *Cupressus lusitanica* productivity in New Zealand. *Forest Ecology and Management, 258*(3), 217–223.

44. Watt, PJ (2005). An evaluation of LiDAR and optical satellite data for the measurement of structural attributes in British upland conifer plantation forestry. England: Department of Geography, University of Durham.
45. Watt, PJ, & Watt, MS (2011). Applying satellite imagery for forest planning. *New Zealand Journal of Forestry, 56*, 23–25.
46. Watt, PJ, & Watt, MS (2012). Forest planning applications using high resolution satellite data. *New Zealand Journal of Forestry, 57*, 32–40.
47. Yu, XW, Hyyppa, J, Kaartinen, H, Maltamo, M (2004). Automatic detection of harvested trees and determination of forest growth using airborne laser scanning.*Remote Sensing Of Environment, 90*(4), 451–462.
48. Zhao, K, & Popescu, S (2009). Lidar-based mapping of leaf area index and its use for validating GLOBCARBON satellite LAI product in a temperate forest of the southern USA. *Remote Sensing Of Environment, 113*, 1628–1645.
49. Zheng, DL, Rademacher, J, Chen, JQ, Crow, T, Bresee, M, le Moine, J, et al. (2004). Estimating aboveground biomass using landsat 7 ETM+ data across a managed landscape in northern Wisconsin, USA. *Remote Sensing Of Environment, 93*(3), 402–411.

Citations

CHAPTER 1

Ethan Warner, Daniel Inman, Benjamin Kunstman, Brian Bush, Laura Vimmerstedt, Steve Peterson, Jordan Macknick, and Yimin Zhang, Modeling Biofuel Expansion Effects on Land Use Change Dynamics, doi:10.1088/1748-9326/8/1/015003.

CHAPTER 2

Andrew D. Barnes, Malte Jochum, Steffen Mumme, Noor Farikhah Haneda, Achmad Farajallah, Tri Heru Widarto, and Ulrich Brose, Consequences of Tropical Land Use for Multitrophic Biodiversity and Ecosystem Functioning, doi:10.1038/ncomms6351.

CHAPTER 3

Feng Wu, Jinyan Zhan, Qian Xu, Yihui Xiong, and Zhongxiao Sun, A Comparison of Two Land Use Simulation Models under the RCP4.5 Scenario in China, doi:10.1155/2013/578350.

CHAPTER 4

Abdullah A. Jaradat and George Boody, "Modeling Agro ecosystem Services under Simulated Climate and Land-Use Changes," ISRN Ecology, vol. 2011, Article ID 568723, 17 pages, 2011. doi:10.5402/2011/568723.

CHAPTER 5

Mohsen Forouzangohar, Neville D. Crossman, Richard J. MacEwan, D. Dugal Wallace, and Lauren T. Bennett, "Ecosystem Services in Agricultural Landscapes: A Spatially Explicit Approach to Support Sustainable Soil Management,", doi:10.1155/2014/483298.

CHAPTER 6

Marie J Heaphy*, David J Lowe, David J Palmer, Haydon S Jones, Gerty JHP Gielen,Graeme R Oliver and Stephen H Pearce, Assessing Drivers of Plantation Forest Productivity on Eroded And Non-Eroded Soils in Hilly Land, Eastern North Island, NewZealand, doi:10.1186/s40490-014-0024-5.

CHAPTER 7

Christine Rösch, Cisco Aust, and Juliane Jörissen, Envisioning the sustainability of the production of short rotation coppice on grassland, doi: 10.1186/2192-0567-3-7.

CHAPTER 8

Jon I Barnes, James MacGregor, and Moira Alberts, Expected Climate Change Impacts on Land and Natural Resource Use in Namibia: Exploring Economically Efficient Responses, doi:10.1186/2041-7136-2-22.

CHAPTER 9

Hartmut Meyer, Systemic risks of genetically modified crops: the need for new approaches to risk assessment, doi:10.1186/2190-4715-23-7.

CHAPTER 10

Pete Watt, Andrew Meredith, Chen Yang, and Michael S Watt, Development of Regional Models of Pinus Radiate Height from GIS Spatial Data Supported with Supplementary Satellite Imagery, doi:10.1186/1179-5395-43-11.

Index